AF411412

Assessment, Diagnosis and Service Life Prediction

Assessment, Diagnosis and Service Life Prediction

Editor

Ana Silva

MDPI • Basel • Beijing • Wuhan • Barcelona • Belgrade • Manchester • Tokyo • Cluj • Tianjin

Editor
Ana Silva
CERIS, Instituto Superior
Técnico, University of Lisbon
Portugal

Editorial Office
MDPI
St. Alban-Anlage 66
4052 Basel, Switzerland

This is a reprint of articles from the Special Issue published online in the open access journal *Buildings* (ISSN 2075-5309) (available at: https://www.mdpi.com/journal/buildings/special_issues/Assessment_Diagnosis_Service_Life_Prediction).

For citation purposes, cite each article independently as indicated on the article page online and as indicated below:

LastName, A.A.; LastName, B.B.; LastName, C.C. Article Title. *Journal Name* **Year**, *Volume Number*, Page Range.

ISBN 978-3-0365-6235-3 (Hbk)
ISBN 978-3-0365-6236-0 (PDF)

Cover image courtesy of Ana Silva

Contents

About the Editor

Ana Silva

Ana Silva is the co-author of 74 scientific publications related to the service life prediction and maintenance of building components. She is co-author of the International Book entitled "Methodologies for service life prediction of buildings: with a focus on façade claddings". She is Responsible Investigator of the Research Project "Buildings' Envelope SLP-based Maintenance: reducing the risks and costs for owners" and Member of the Research Projects "Development and assessment of models for the durability and preventive conservation of historic plasterwork from the decorative elements of the Real Alcázar of Seville" and "Inteligencia artificial al servicio de la gestión integral de inmuebles de conservación histórica en el sur de Chile: la región de los ríos y la región de los Lagos". She is secretary of the CIB W080 commission of the International Council for Research and Innovation in Building and Construction. She has supervised ten master theses and four PhD theses and has participated in 52 scientific events.

 *buildings*

Editorial

Assessment, Diagnosis and Service Life Prediction

Ana Silva

Civil Engineering Research and Innovation for Sustainability (CERIS), Department of Civil Engineering, Architecture and Georesources, Instituto Superior Técnico, University of Lisbon, Av. Rovisco Pais, 1049-001 Lisbon, Portugal; ana.ferreira.silva@tecnico.ulisboa.pt

Citation: Silva, A. Assessment, Diagnosis and Service Life Prediction. *Buildings* **2022**, *12*, 2005. https://doi.org/10.3390/buildings12112005

Received: 12 October 2022
Accepted: 15 November 2022
Published: 17 November 2022

Publisher's Note: MDPI stays neutral with regard to jurisdictional claims in published maps and institutional affiliations.

Service life prediction is crucial for the adoption of more sustainable solutions, allowing optimizing the costs and environmental impact of buildings during their life cycle. Accurate assessment of the service life of buildings requires a thorough understanding of degradation mechanisms and materials' behaviour. Building pathology assessment methods allow characterizing the deterioration state of the buildings and their components, using as indicators specific measurable properties. Based on this information, different service life prediction methodologies can be defined in order to provide reliable data concerning the most probable failure time of buildings and components according to their characteristics and their age.

This Special Issue intends to provide new perspectives on the existing knowledge related to various aspects of *Assessment, Diagnosis and Service Life Prediction* of buildings and components. The ten original research studies published in this Special Issue come from research centres and university departments of Civil and Construction Engineering, Safety Management, Building and Real Estate, Environmental Engineering, Sustainability and Innovation in Structural Engineering, Geotechnical Engineering, Architecture and Built Environment, with the relevant contribution of international experts from Australia, Brazil, Czech Republic, Hong Kong, Iran, Israel, Norway, Portugal and Taiwan.

All constructions endure a gradual degradation process, from the moment they are built and put into use [1], due to the action of several degradation agents, such as solar radiation, high temperatures, wind-driven rain, freeze–thaw cycles, the presence of chemical substances, among others [2,3]. Therefore, the assessment of the degradation condition of constructions over time is crucial for an adequate diagnosis, service life prediction and management of the maintenance activities over the years. Different approaches can be adopted to model the degradation mechanisms over the constructions' lifetime, namely, (i) fieldwork surveys and (ii) laboratory testing.

Fieldwork surveys, based on periodic inspections, allow obtaining relevant information about the defects that may occur throughout the constructions' service life [4]. In this approach, the degradation condition of constructions when subjected to real exposure conditions is evaluated adopting an ex post facto approach, i.e., making an anamnesis of the construction by studying possible relationships between the anomalies observed and probable causes. Adopting this approach, Coelho et al. [5] proposes a service life prediction method, applied to wood flooring systems, based on the collection of the degradation condition of 96 indoor wood floorings' in-use conditions, located in Portugal. This study evaluates the impact of various characteristics that influence the floors' durability, showing the high importance of the type of protection, the type of wood, and the type of floor on the estimated service life of wood floorings. Mousavi et al. [6] adopt a similar approach, proposing a degradation model based on an extensive fieldwork survey to 162 stone claddings directly adhered to the substrate, located in Tehran. The authors [6] obtained an estimated service life of 65 years, concluding that the exposure to environmental agents, such as wind, rain, and pollutants, is the main cause of degradation of the natural stone claddings in Tehran. Additionally, using visual inspections and archival research, Kroftova [7] identify the most common defects and failures of masonry structures in urban residential buildings from the second half of the 19th century and the start of the 20th century. This study [7] highlights that a detailed and accurate analysis of the causes and consequences of defects is a crucial precondition for the reliability and long-term durability of rehabilitation of buildings.

Usually, visual inspections are the most common technique to assess the degradation condition of constructions, and, in most cases, the information collected in this way is appropriate to support the decision to intervene [8]. Nevertheless, new techniques have emerged that allow improving and automatizing the collection of reliable on-site data, reducing the uncertainty of the diagnosis. Dias et al. [8] provide an overview of the emerging technologies for inspection (e.g., 3D laser scanning, infrared thermography, photogrammetry, digital image processing, and drones) and perform a critical analysis of the most adequate technique to detect a specific anomaly in buildings' façades.

Real life conditions and the synergy between different degradation agents are difficult to model in a laboratory [9]. Even so, laboratory tests allow obtaining results more quickly and properly modeling the effects of a specific degradation agent. Asphaug et al. [3] investigate the risk of moisture accumulation in a multi-layered concrete façade system exposed to accelerated ageing in a climate simulator according to NT Build 495 [10]. This study [3] reveals that the amount of moisture accumulation depends strongly on the type of concrete and whether a water-repellent surface treatment is applied, providing relevant knowledge about the degradation condition and thermal performance of the façade systems under analysis.

The reference service life of a construction is often based on historical data, previous experience or the performance of the materials analyzed under similar conditions. Ingebretsen et al. [11] perform a scoping study on the current body of knowledge on the microclimate of air cavities in ventilated roofing and claddings in Nordic climates. This study [11] reveals that the existing knowledge about climatic conditions in air cavities in ventilated building roof and façade systems is limited. The design of future structures only based on experience is not adequate, not allowing anticipating the impacts of future climate changes on the risk of rot and mold growth in air cavities.

Degradation models and service life prediction are also crucial for evaluating the performance of new materials and building systems. John et al. [12] present an experimental study for the creation of a novel modular flooring system that can be flat-packed and built into modular housing components on-site, answering to an increasing demand for lightweight modular construction. The assessment of the new system condition after failure mode tests reveals an adequate structural performance and ease in fabrication as opposed to the conventional formworks and commercial temporary flooring systems.

The assessment, inspection and diagnosis of the construction and building materials are crucial to maintain them in adequate performance conditions over time. The adoption of adequate maintenance is the most cost-effective way to minimize the constructions' deterioration while extending their service life. Nevertheless, even with a rational maintenance plan, there is always a risk that the constructions deteriorate at a faster pace, compared to what was planned, requiring the anticipation of the necessary maintenance actions, so as not to compromise the durability of these elements. To overcome the existing gap, Dias et al. [13] propose property maintenance insurance policies developed based on condition-based maintenance plans, to face the risk of needing anticipated maintenance in buildings' façades. This insurance policy allows changing the nature of the risk and its allocation, transferring the risk to the insurer, and increasing the asset's equity value, reducing the risk associated with the degradation of buildings and components. Santos et al. [4] analyses the risks of not considering distinct structural deterioration processes over time depending on their age, location, structural type, and other aspects, on the definition of bridge management systems. This study [4] affirms that a superior level of safety and serviceability must be reached to ensure the operating status of a bridge network, which is crucial for countries' social and economic development. In this sense, the authors [4] propose degradation models to predict bridge deterioration and improve the inspection periodicity, improving the reliability of the decision process of the bridge management system (BMS) in Brazil. The study by Wang et al. [14] emphasizes the synergy between the maintenance conditions and the safety of educational institutions and public facilities. This study [14] proposes a robust framework for a risk-informed integrated safety–maintenance management framework for educational and public facilities, revealing that maintenance activities significantly affect the safety of electric system components, infrastructures, fire protection systems and structural components. The results of this study [14] suggest that annual safety audits of the systems seem to be insufficient, and higher frequencies of maintenance as well as safety audits with intervals of between 3 and 6 months are suggested.

Now more than ever, the construction industry must adopt more durable solutions and rational decisions at the design and maintenance stages. The different papers published in this Special Issue address the strategic and economic relevance of assessment, diagnosis, and service life prediction of constructions, ranging from facades or components of residential buildings to critical buildings such as schools or bridges. The knowledge about the degradation processes and service life prediction is crucial, considering a detailed analysis of the causes and consequences of defects. An important message from this Special Issue is that the future studies must propose maintenance management frameworks based on the knowledge of the degradation pattern of constructions, adopting proper maintenance actions, in adequate timeframes, considering the specific conditions of the construction under analysis, to ensure the functionality, durability and safety of the construction, while minimizing the financial investment. Moreover, most of the existing studies adopt standard values for the buildings' service life (e.g., 50 years) when performing life cycle cost (LCC) and life cycle assessment (LCA) analyses. In reality, the service life of a building varies according to the materials applied, the construction methods and design, the exposure to environmental conditions, and in-use and maintenance conditions. The adoption of standard assumptions about the service life of buildings and components may lead to incorrect results, leading to rough estimations of their real environmental impact. In this sense, service life prediction and the knowledge about the durability of buildings and their components has a crucial role in accurately assessing their sustainability or environmental impacts over the years.

The editor would like to acknowledge the generosity of all the authors who shared their scientific knowledge and expertise in different fields related to *Assessment, Diagnosis and Service Life Prediction*. Moreover, the editor would like to express their gratitude to the peer reviewers for their rigorous analysis of the different contributions, who have appreciably contributed to enriching the quality of this Special Issue, and finally, to the managing editors of *Buildings*, who have continuously supported everyone involved in this Special Issue.

Funding: This research received no external funding.

Conflicts of Interest: The author declares no conflict of interest.

References

1. Silva, A.; de Brito, J.; Gaspar, P.L. *Methodologies for Service Life Prediction of Buildings: With a Focus on Façade Claddings*, 1st ed.; Springer: Geneva, Switzerland, 2016.
2. Jelle, B.P. Accelerated climate ageing of building materials, components and structures in the laboratory. *J. Mater. Sci.* **2012**, *47*, 6475–6496. [CrossRef]
3. Asphaug, S.K.; Time, B.; Kvande, T. Moisture Accumulation in Building Façades Exposed to Accelerated Artificial Climatic Ageing—A Complementary Analysis to NT Build 495. *Buildings* **2021**, *11*, 568. [CrossRef]
4. Santos, A.F.; Bonatte, M.S.; Sousa, H.S.; Bittencourt, T.N.; Matos, J.C. Improvement of the Inspection Interval of Highway Bridges through Predictive Models of Deterioration. *Buildings* **2022**, *12*, 124. [CrossRef]
5. Coelho, P.; Silva, A.; de Brito, J. How Long Can a Wood Flooring System Last? *Buildings* **2021**, *11*, 23. [CrossRef]
6. Mousavi, S.H.; Silva, A.; de Brito, J.; Ekhlassi, A.; Hosseini, S.B. Degradation Assessment of Natural Stone Claddings over Their Service Life: Comparison between Tehran (Iran) and Lisbon (Portugal). *Buildings* **2021**, *11*, 438. [CrossRef]
7. Kroftova, K. Most Frequent Problems of Building Structures of Urban Apartment Buildings from 2nd Half of 19th Century and the Start of 20th Century. *Buildings* **2021**, *11*, 27. [CrossRef]

8. Dias, I.S.; Flores-Colen, I.; Silva, A. Critical Analysis about Emerging Technologies for Building's Façade Inspection. *Buildings* **2021**, *11*, 53. [CrossRef]
9. Kus, H.; Carlsson, T. Microstructural investigations of naturally and artificially weathered autoclaved aerated concrete. *Cem. Concr. Res.* **2003**, *33*, 1423–1432. [CrossRef]
10. NT Build 495. *Building Materials and Components in the Vertical Position: Exposure to Accelerated Climatic Strains*; Nordtest: Espoo, Finland, 2000.
11. Ingebretsen, S.B.; Andenæs, E.; Kvande, T. Microclimate of Air Cavities in Ventilated Roof and Façade Systems in Nordic Climates. *Buildings* **2022**, *12*, 683. [CrossRef]
12. John, K.; Rahman, S.; Kafle, B.; Weiss, M.; Hansen, K.; Elchalakani, M.; Udawatta, N.; Hosseini, M.R.; Al-Ameri, R. Structural Performance Assessment of Innovative Hollow Cellular Panels for Modular Flooring System. *Buildings* **2022**, *12*, 57. [CrossRef]
13. Dias, I.S.; Silva, A.; Cruz, C.O.; Ferreira, C.; Flores-Colen, I.; de Brito, J. Insurance Policies for Condition-Based Maintenance Plans of ETICS. *Buildings* **2022**, *12*, 707. [CrossRef]
14. Wang, K.-C.; Almassy, R.; Wei, H.-H.; Shohet, I.M. Integrated Building Maintenance and Safety Framework: Educational and Public Facilities Case Study. *Buildings* **2022**, *12*, 770. [CrossRef]

Article

Moisture Accumulation in Building Façades Exposed to Accelerated Artificial Climatic Ageing—A Complementary Analysis to NT Build 495

Silje Kathrin Asphaug [1,2,*], **Berit Time** [2] **and Tore Kvande** [1]

[1] Department of Civil and Environmental Engineering, Norwegian University of Science and Technology, NO 7491 Trondheim, Norway; tore.kvande@ntnu.no
[2] SINTEF Community, Department of Architecture, Materials and Structures, NO 7465 Trondheim, Norway; berit.time@sintef.no
* Correspondence: silje.asphaug@sintef.no

Abstract: Building façades must endure severe climatic exposure throughout their lifetimes. To prevent damage and expensive repairs, ageing tests are used in durability assessments. The NT Build 495 describes an artificial ageing procedure to address building material and component resistance to ultraviolet (UV) light, heat, water, and frost using a climate simulator. The test has been used for decades to investigate exterior surface materials and façade products but has only recently been adopted for multi-layered systems. This study investigates moisture accumulation in a façade system for retrofitting based on concrete and thermal insulation. Hygrothermal simulations of the façade system subjected to ageing were conducted. Moisture accumulation was considered theoretically for the current test procedure and compared to a modified setup in which the interior climate was controlled at 21 °C. Physical measurements were performed in the climate simulator to determine the boundary conditions. Results showed that moisture accumulation in the thermal insulation was largely affected by the type of concrete, that applying a water-repellent surface treatment reduced moisture accumulation, and that the current setup resulted in less moisture accumulation compared to the modified setup. The latter implicates accelerated degradation with the modified setup.

Keywords: building defects; accelerated ageing; durability; ETICS; moisture control; climate adaptation; mass transfer properties; hygrothermal simulations

Citation: Asphaug, S.K.; Time, B.; Kvande, T. Moisture Accumulation in Building Façades Exposed to Accelerated Artificial Climatic Ageing—A Complementary Analysis to NT Build 495. *Buildings* **2021**, *11*, 568. https://doi.org/10.3390/buildings11120568

Academic Editor: Ana Silva

Received: 19 October 2021
Accepted: 18 November 2021
Published: 23 November 2021

Publisher's Note: MDPI stays neutral with regard to jurisdictional claims in published maps and institutional affiliations.

1. Introduction

Buildings are exposed to severe degrading factors throughout their lifetime, including solar radiation, high temperatures, wind-driven rain, freeze–thaw, and chemical substances [1]. More intense rainfall events and increased annual precipitation are predicted in many geographical regions with cold climates owing to climate change [2]. When a building's "weather skin" fails to withstand climatic exposures over time, failures/leaks occur. Moisture penetrating the building envelope can lead to a range of damages, influence occupants' health and comfort, incur extensive costs, and reduce the thermal performance of the envelope [3–6]. As damage can be prevented by selecting durable materials and façade systems, accelerated ageing tests have emerged as an important tool to promote sustainable buildings [7].

Accelerated artificial ageing involves subjecting materials to climatic factors similar to those experienced through a building's service life, but at a high intensity over a shorter period. In this way, the vulnerability of building products or façade systems can be tested within a reasonable time. Many different types of climate ageing laboratory apparatuses have been developed for this purpose [1], including the climate simulator described by Nordtest NT Build 495 [8]. The method is intended to expose the materials and components used in a building envelope to ultraviolet (UV) light, heat, water, and frost. During the accelerated artificial climatic ageing test NT Build 495, the specimens are positioned vertically in a rotating carousel and exposed to the four degrading factors repeatedly for one hour each (see Section 2.2). The method does not include resistance to mechanical loads such as hail, which might be an issue for Exterior Thermal Insulation Composite Systems (ETICS) [9]. Other relevant degrading factors omitted from the test are environmental pollution and dirt deposition, mould growth, vibrations, and wind [3].

The SINTEF, as the leading test facility in Norway, has used the NT Build 495 climate simulator as a standardised exposure method for building materials and components for several decades. Comparing test results to experiences from field investigations has contributed to substantiating the applicability of the test method [1,6]. The test is primarily used to consider whether materials and components can withstand climatic exposures as harsh as those they may be subjected to in real life; if the exposure in the simulator is endured, the durability is likely to be sufficient. Improvements are called for if damage occurs after a short exposure time. Through repetitive testing, improvements to the material composition or component design can be investigated and compared with the initial results. Wind barriers, wood shingles and façade claddings, underlayer roofs, tape for building purposes, brick, sealings, windows, and renders are examples of façade materials which through testing according to NT Build 495 have achieved a documented level of durability [10–13]. The test is also applied to façade systems such as ETICS because the outer surface is vulnerable to cracking owing to freeze–thaw [6] and elevated temperatures caused by solar radiation [14]. When exposed to severe driving-rain conditions, assemblies with only a single-stage protection against wind and rain are particularly vulnerable, as cracking of the exterior surface leads to leaks. Owing to the limited drying ability, even small cracks can cause moisture to accumulate in the underlying structure and lead to damage [6].

Rainwater may also be transferred into the underlying structure by capillary conduction or diffusion; the latter may occur when the exterior surface is wetted by rain and subsequently heated by solar radiation. The uptake of rainwater at the surface depends mainly on the capillary properties of the exterior materials, and whether a water-repellent surface treatment is applied. The impact of these transfer mechanisms on the overall moisture performance of assemblies varies depending on the mass transfer properties which are often lacking, time-consuming, or expensive to measure, and may show large standard deviations for the same material [15]. Numerical simulations are commonly used to investigate the drying ability of assemblies; however, results are often prone to large deviations because of large uncertainties in the mass transfer properties and the inability to realistically include the diffusion and suction contacts between material layers [15].

In Norway, Technical Approvals authorized by SINTEF Certification, document to builders that a façade system is suitable for the Norwegian climate. The criteria to obtain a Technical Approval are mainly based on methods and criteria for assessing the performance given by EAD 040083-00-0404 [16], but testing of durability properties according to NT Build 495 is additionally required. The NT Build 495 has been found to be an applicable test method in Norway [6]; however, the test is currently carried out without a habitable climate on the indoor side of the specimen. Façade systems are subjected to degrading climatic factors on the exterior side and the temperature/relative humidity (RH) on the interior side is currently not controlled; that is, the temperature/RH on the interior side follows the fluctuations on the exterior side. Although this practice works adequately in most cases, it is necessary to investigate whether the interior climate should be controlled when moisture-sensitive assemblies are assessed—e.g., assemblies with only single-stage protection against rain intrusion.

In the authors' opinion, omitting to control the interior climate (as in the current test setup) should not result in more moisture accumulation compared to a modified setup where the interior parts of the façade system are included, and the interior climate is controlled. To consider the need to change the current test methodology of NT Build 495 to control the interior climate, a complementary numerical analysis has been performed. This study investigates how moisture accumulation in a façade system exposed to accelerated artificial ageing is affected by the interior climate.

To address these general inquiries and uncertainties, the following research questions were raised:

1. To what extent does moisture accumulate in a façade system with single-stage weather protection when exposed to climate exposure in the climate simulator?
2. How does the interior climate in the climate simulator affect moisture accumulation in façade systems with single-stage protection?

Given the extent of the research field, certain limitations were determined:

1. This study concerns a façade system with single-stage protection against rain and wind. Assemblies with two-stage protection were not addressed.
2. In the hygrothermal simulations, we presumed that the façade system endured the accelerated ageing test without cracking of the exterior surface and that the mass transfer properties did not change over time owing to degradation.
3. Material properties (including mass transfer properties) were mainly obtained from the WUFI®Pro material database [17]; hence, they were not measured. The exception is the material properties of the polyisocyanurate-insulation (PIR insulation), which was defined based on input from the producer of the façade system.
4. The moisture content in the façade system or the degree of degradation occurring throughout the ageing test were not measured.
5. Hysteresis in the concrete moisture storage function was not accounted for in the numerical simulations.
6. Changes in material properties caused by degradation or chemical processes, such as curing, were not accounted for.
7. The simulations were not validated with physical measurements.

2. Theoretical Framework

2.1. Durability of Façade Systems

Façade systems are primarily subjected to accelerated ageing to assess the durability of an exterior surface because the ability to withstand UV light, heat, water, and frost are crucial to a building's climate shell/weather skin. However, all adjoining materials must collectively withstand climatic exposures to achieve the best possible durability.

Façade systems erected in accordance with the principle of two-stage tightening is considered a robust solution; that is, weather protection by a separate rain shield (cladding) and air/wind tightening (wind barrier). Two-stage tightening is crucial in climates with severe driving-rain exposure [18] as in large parts of the west coast of Norway [6]. The risk of moisture-related damage decreases when moisture penetrating the exterior surface (the rain shield) is allowed to dry effectively through a ventilated air gap. Façade systems with only single-stage protection against wind and precipitation (e.g., ETICS) are much more vulnerable to damage because cracking of the exterior surface can lead to leaks. Due to their limited ability to dry, even small cracks can cause moisture to accumulate in the underlying structure and lead to damage [6]. Therefore, the durability of the exterior surfaces of façade systems with only single-stage protection is of major importance. A known cause of failure is the stress that occurs when adjoining materials expand and contract differently owing to temperature changes [19,20]. For ETICS, the critical factors in this context are the stiffness and thermal expansion coefficient of the thermal insulation along with the thermal expansion coefficient of the exterior rendering. Restrained movements of the components can result in bending moments, particularly when the insulation material has a high stiffness value [21]. Other major causes of degradation are frost weathering during water-to-ice volume expansion, photodegradation caused by solar radiation, and hail [9]. These and several other climate exposure factors were further addressed by [1].

Considering only natural weather ageing to gain knowledge of the durability of materials or components is often unsatisfactory because the natural outdoor climate ageing processes take time [1]. Accelerated climate ageing tests are an option to obtain sufficiently fast and economical results. Accelerated ageing is considered a valuable practice to avoid extensive building damage owing to missing or too low resistance to climate exposure [1]. A number of accelerated climate ageing apparatuses exist and can be utilised in the laboratory according to different test methods and standards [1,22]. The purpose of accelerated climatic ageing tests is, however, not to provide an accurate estimate of service life expressed in terms of a number of years; it is to benchmark the ageing properties of different materials, systems, and techniques [6].

2.2. Accelerated Artificial Ageing of Façade Systems

To assess the durability of façade systems/assemblies, test equipment further referred to as the climate simulator is much-applied. The accelerated artificial climatic ageing in the climate simulator is described by Nordtest NT Build 495 [8]. The test was developed 50 years ago by Byggforsk (which later merged with SINTEF). After many years of experience including comparisons with natural climatic ageing of façade materials, Byggforsk developed a method description which was later approved as the Nordtest method [8].

During the accelerated artificial climatic ageing test NT Build 495, specimens are positioned vertically in a rotating carousel and exposed to four degrading factors repeatedly for 1 h each (see Figure 1). In the first chamber, UV radiation is applied using fluorescent UV tubes with a relative spectral distribution in the UV band close to that of global solar irradiance. The black panel temperature rises to its designated temperature (normally $63 \pm 5\,°C$) in 45 min from the beginning of exposure to UV light and heat radiation. If required, the black panel temperature may be chosen as $35 \pm 5\,°C$, $50 \pm 5\,°C$, or $75 \pm 5\,°C$, cf. ISO 4892. The temperature is controlled using infrared halogen lamps. In the second chamber, the specimens are wetted with a spray of demineralised water. The suggested strain was $15 \pm 2\,L/m^2 \cdot h$, but various spraying conditions may be used if required. To allow water to drip off, the spraying is terminated 10 min before the specimens rotates in to the third chamber. In the third chamber, an air temperature of $-20 \pm 5\,°C$ is suggested, but other specified air temperatures may be used if registered and reported. In the fourth chamber, the specimens are thawed at a laboratory climate of $23 \pm 5\,°C$ and $50 \pm 10\%$ RH.

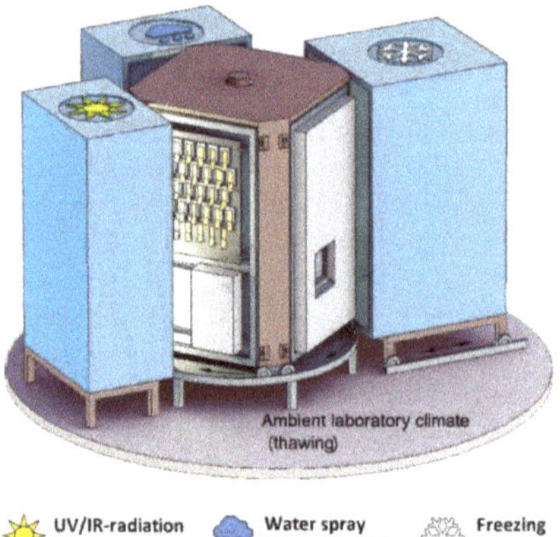

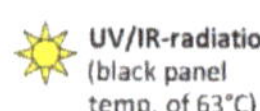

Figure 1. Climate simulator with four climate zones used for testing according to NT Build 495 as illustrated by Kvande et al. [6]. The mid-section is rotating clockwise.

The results of the test include information on the changes occurring during the test, the scale of the changes, and the time of occurrence. The results are assessed qualitatively, and the performance properties is often a change in performance properties—change in appearance of the specimens during the test or signs of degradation; for example, cracks, loss of gloss, or delamination.

Testing of façade solutions according to NT Build 495 [8] is currently carried out without the interior parts of the walls and without a habitable climate on the indoor side of the specimen. It is uncertain how this affects the moisture transfer at the surface and the moisture performance of the tested specimens.

Although this might be a shortcoming in the current test setup, it is a setup much-applied to assess the durability of exterior surfaces. E.g. D'Orazio et al. [23] investigated the durability of an EPS-based lightweight prefabricated construction system without focusing on the impact of the interior climate. Griciutė et al. [22] compares water adsorption of ETICS samples with samples on plaster strips but does not focus on the interior parts of the wall or the interior climate. Franzoni et al. [24] performed accelerated ageing tests to investigate the durability of a new prefabricated external thermal insulation composite board for building retrofitting without focus on the interior parts of the walls nor the indoor climate. In addition there are many examples of studies focusing on accelerated artificial ageing of small samples, such as [25,26].

Daniotti et al. [3] on the other hand, assessed the degradation evolution and the loss in hygrothermal performance of ETICS using a test setup which included all layers of the wall and a laboratory climate on the indoor side. They observed decreasing thermal resistance due to an increase of water content caused by rain penetration. They concluded that ageing, and moisture, whose content within layers is increased by ageing, affect dynamic thermal performances. The main degrading factor highlighted in this study, however, seems to be the thermal shock and dilatation-contraction events which causes blistering and deformation of the top coat. Gonçalves et al. [14] assessed the degradation and hygrothermal performance of the vacuum-insulation panel (VIP) based ETICS using a test setup which included all layers of the wall and a laboratory climate on the indoor side. They compared the test procedures defined in [16] with a new procedure to simulate solar radiation conditions. The new procedure caused defects such as loss of flatness and finishing coat microcracking, which were not found after the standard procedures, revealing the importance of studying the radiation effect on ETICS systems.

Other noteworthy studies need also be mentioned. E.g. Maia et al. [27] investigated the hygrothermal performance of a new thermal aerogel-based render applied as a component of a multilayer coating system by measuring relevant material properties and conducting hygrothermal simulations. The study highlights the importance of applying a finishing coating with low capillary absorption to reduce the water content in the inner layer and consequently the impact on the U-value of the façade. The numerical simulations highlighted that the hygrothermal risk increased in more severe climates (such as Hannover in Germany compared to Porto in Portugal) but does not focus on colder climates such as Norway. The importance of taking into account freeze–thawing ageing in colder and moderate climates is however assessed by Maia et al. [28].

2.3. Moisture Transfer in Façade Systems with Single-Stage Protection

Façade systems with only single-stage protection against wind and rain are considered particularly vulnerable to moisture damage because cracking of the exterior surface can lead to leaks and cause moisture to accumulate in the structure. Moisture from the exterior surface might also be transferred inward to the thermal insulation by diffusion when a wet surface is subjected to solar radiation. Without a ventilated air gap behind the exterior surface to enable moisture to dry effectively, moisture might accumulate in the thermal insulation behind the outer layer(s) and cause damage.

Façade systems tested in the climate simulator [8], are subjected to the degrading climatic factors on the exterior side, since the purpose of the test is to detect material degradation caused by exterior climatic exposures. The climate on the interior side is currently not controlled, as it is not within the purpose of the test to assess the overall heat resistance, general moisture transfer or durability of the interior parts of the structure. However, because the inward diffusion of moisture (caused by temperature differences) might be affected by the interior temperature, it might be necessary to control the interior climate when façade systems with only single-stage protection are tested.

In this study, we sought to evaluate the current test setup to determine whether the interior climate should be controlled. Omitting to control the interior climate (as in the current test setup) should not lead to more moisture accumulation compared to a modified test setup where the interior part of the wall is included and the indoor temperature is controlled.

2.4. Hygrothermal Simulations of Façade Systems Subjected to Accelerated Ageing

The ageing and degradation of materials and components cannot be meaningfully investigated by means of numerical simulations only [3]. To gain information on the durability, testing by either natural ageing or accelerated artificial climatic ageing must be carried out [1]. In this study, however, the objective was to investigate the moisture transfer from the exterior surface (exposed side) through the concrete and into the thermal insulation to determine whether it is necessary to control the interior climate in the climate simulator. As a theoretical approach, we assumed that the façade system was subjected to accelerated ageing for 12 months without cracking of the exterior surface and without significant changes to the hygrothermal properties. In this way, we may investigate the possible moisture accumulation within the thermal insulation, not exclusively caused by cracking of the exterior surface, and determine how the moisture transfer is affected by the interior climate.

2.5. Calculating the Capillary Absorption of Rain

To calculate the capillary absorption of rain, WUFI®Pro [17] uses two material properties: D_{ws} (moisture diffusivity coefficient for "suction" (m^2/s)) and D_{ww} (moisture diffusivity coefficient for redistribution (m^2/s)). The D_{ws} represents the capillary uptake of water when the surface of a sample element is completely wetted such as during a rain event. The suction transport is dominated by the larger capillaries because their lower capillary tension is more than compensated for by their markedly lower flow resistance. The D_{ww} describes the dispersal of absorbed water when the wetting is complete; that is, when no more water is taken up and the water in the material is further distributed. For a building component, this corresponds to moisture transport when rain is absent. The redistribution is then dominated by the smaller capillaries because their higher capillary tension draws the water out of the larger capillaries (WUFI®Pro online help [17,29]). Moisture absorption from rain is neglected in WUFI®Pro if the "adhering fraction of rain" is set to 0 or if the rainfall is lower than 0.1 mm (it is then assumed that the rain evaporates before it is drawn into the surface). Rain is also used to determine whether capillary absorption (D_{ws}) or capillary redistribution (D_{ww}) is dominant in a building component. That is, for time steps where the climate data show rain, D_{ws} will be used to calculate the capillary transport processes in the component. When rain is absent or very low, D_{ww} is used. The D_{ws} is generally significantly larger than D_{ww} because redistribution is a slower process that occurs in small capillaries with high flow resistance (WUFI®Pro online help [17,29]). It is uncertain how well suited the software is to realistically replicate the uptake and distribution of rain in a multi-layered façade system with a concrete surface repeatedly subjected to rain events for a long period of time.

3. Methodology

Hygrothermal simulations were performed for a façade system subjected to accelerated artificial climatic ageing in the climate simulator. Simulations were performed according to the test setup currently used in the climate simulator and compared to a modified test setup in which the interior climate was controlled. The façade system was simulated with and without a water-repellent surface treatment on the exterior side, and five different types of concrete were compared. We assumed that the façade system was subjected to accelerated ageing for 12 months without cracking of the exterior surface and without significant changes to the hygrothermal properties. Measurements were performed in the climate simulator to determine the interior and exterior boundary conditions.

3.1. Overall Approach

The façade system chosen for this investigation is intended to be used on the exterior side of the existing timber frame walls as a part of façade refurbishment. The façade system consists of two 15-mm thick concrete layers with 30-mm PIR-insulation between them and a vapour permeable rigid mineral wool board on the back as shown in Figure 2. The system is of particular interest for this study because it is a commercial Norwegian product, it contains many layers of material and features only single-stage protection. The façade system is also previously tested according to NT Build 495 [8].

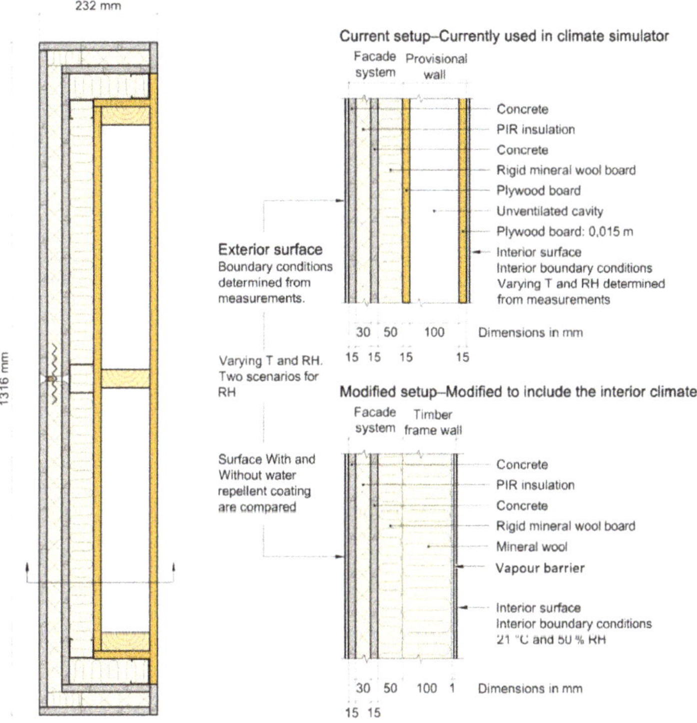

Figure 2. Test setups applied in the hygrothermal simulations. In the current setup, the façade system was erected on a provisional wall and the interior climate varied depending on the exterior exposure. In the modified setup, the provisional wall was replaced with a mineral wool board and an interior vapour barrier. The exterior boundary conditions were the same as those of the current setup, but a constant temperature and RH were set at the interior boundary.

Hygrothermal simulations were performed using the commercial software WUFI®Pro version 6.4 [17]. Due to the fact that the simulation program addresses one-dimensional heat and moisture transport, 3D effects from frameworks, steel and hat profiles, fasteners, joints, and edges were neglected. The simulations were performed for two different test setups (i.e., the current setup and the modified setup) (see Figure 2).

The current setup constituted the test setup currently used to carry out accelerated ageing tests in the climate simulator; the façade systems are mounted in the climate simulator on a provisional wall consisting of two plywood boards with an air cavity in between. The exterior and interior boundary conditions were determined from measurements in the climate simulator. The interior climate was therefore not controlled but followed fluctuations in temperature and RH from exposure on the exterior side.

The modified setup constituted a hypothetical situation in which the interior climate in the climate simulator is controlled at 21 °C/50% RH. The objective of this kind of modification is to create a climate that is more representative of heated buildings and the heat and moisture transfer that occurs within building façades when exposed to a varying exterior climate. In the modified setup, the provisional wall was replaced with an insulated timber frame wall with a vapour barrier on the interior side. The exterior boundary conditions were the same as those in the current setup. For the interior boundary conditions, a constant temperature and RH were chosen.

3.2. Material Properties and Initial Conditions

The hygrothermal properties of the materials used in the façade system were not documented or published by the manufacturer. Due to the fact that the properties of concretes vary significantly, five different types of concrete were compared as shown in Table 1. The concrete and other materials in the façade system were selected from the WUFI®Pro material database. An overview of the material properties is provided in Figure A1 and Table A1 in Appendix A. The initial temperature was set to 5 °C. The typical built-in moisture contents suggested by WUFI®Pro were used as the initial condition as shown in Table A1.

Table 1. Types of concrete used in hygrothermal simulations.

Name of Concrete in This Study	Name in WUFI®Pro Database	Source	Quality of Concrete (WUFI®Pro Material Database)	Colours in Results
Old	Concrete w/c = 0.4	LTH Lund University, Sweden	Material data applies to several years' old concrete and cannot be used to address built-in moisture. Liquid transfer coefficients are not included.	— Old RH-scenario "Worst case" Old "High"
C12/15	Concrete, C12/15	Fraunhofer-IBP	No additional information	— C12/15 RH-scenario "Worst case" C12/15 "High"
MASEA	Concrete	MASEA Database, Germany	No additional information	— C35/45 RH-scenario "Worst case" C35/45 "High"
C35/45	Concrete, C35/45	Fraunhofer-IBP	No additional information	— Masea RH-scenario "Worst case" Masea "High"
Waterproof	Concrete, w/c = 0.5	Fraunhofer-IBP	Moisture permeability of mature concrete. With respect to the overall moisture transmission, this concrete is roughly equivalent to waterproof types of concrete (appropriate thickness and reinforcement required).	— Waterproof RH-scenario "Worst case" Waterproof "High"

3.3. Boundary Conditions

Climate data files were created using the CreateClimateFile.xlsm Ver.: 2.6 supplied with the WUFI®Pro software [17]. Owing to the rapidly changing conditions in the climate simulator, a 1-h time step was not appropriate for the boundary conditions, and shorter time steps were adapted. In this case, 1/10 h (6-min intervals) proved to be an appropriate resolution. The climate data used in the simulations consisted of a 4-h period which was repeated for 12 months.

The same exterior boundary conditions were used in the simulations of both the current and modified test setups. The conditions were determined based on the measurements of temperature and RH performed in SINTEF's artificial climate simulator (see Appendix B). The temperature and RH variations used in the simulations are listed in Table 2 and are depicted in Figure 3. Both test setups were simulated with and without a water-repellent surface treatment on the exterior side; that is, allowing or excluding rain adsorption. The heat and moisture transfer coefficients on the exterior side were set to zero because the boundary conditions were determined from temperatures measured at the surface. The RH was measured only in the airgap behind the test element in the climate simulator. Therefore, two different RH-scenarios, "worst case" and "high", were determined and used as boundary conditions at the exterior surface.

Table 2. Exterior and interior boundary conditions used in the hygrothermal simulations for the current and the modified test setups.

Exterior climate Current and modified test setup	At exterior surface *	T	Temperature measured on exterior surface of test element (red). Solar radiation is therefore not used as a boundary condition. (WUFI uses air temperature and solar radiation to calculate surface temperatures.)		
		RH	RH determined from measurements, (Appendix B)	RH-scenario "worst case" (dark blue)	Sun: RH calculated from the RH measured in air gap behind test element
					Rain: RH set to 100%
					Frost: RH set to 100%
					Lab: RH set to 100%
				RH-scenario "high" (light blue)	Sun: RH calculated from RH measured in air gap behind test element
					Rain: RH set to 100%
					Frost: RH calculated from T/RH measured in air gap behind test element
					Lab: RH reduced to 76% (value between calculated RH (from measured vapour pressure in air gap behind test element) and measured vapour pressure in the laboratory)
		Rain	15 L/m^2·h (adhering fraction of rain = 1)		
Interior climate Current test setup	Indoor air **	T	Temperature as measured in airspace behind test element (yellow)		
		RH	RH as measured in airspace behind test element (green)		
Interior climate Modified test setup	Indoor air **	T	21 °C		
		RH	50%		

* Heat and moisture transfer coefficients = 0 W/m^{2}K (because T/RH was measured at the surface). ** Heat transfer coefficient = 8.0 W/m^{2}K. The corresponding moisture transfer coefficient was determined by WUFI®Pro.

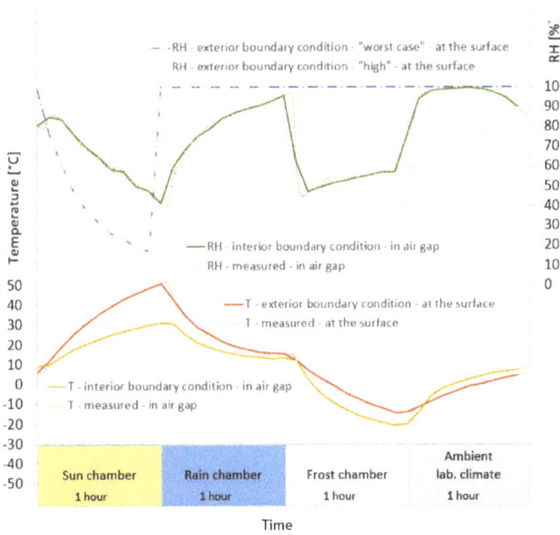

Figure 3. Interior boundary conditions used in the current test setup and exterior boundary conditions used in both the current and the modified test setup. The 4-h period (6-min intervals) corresponds to one rotation in the climate simulator and was repeated for 12 months in the simulations.

The interior boundary conditions used in the simulations of the current test setup were determined from the measurements performed in SINTEF's climate simulator (see Appendix B). The temperature and RH variations are shown in Table 2 and depicted in Figure 3. The heat transfer coefficient was set to 8.0 W/m^{2}K, and the corresponding moisture transfer coefficient was determined by WUFI®Pro. In the simulations of the modified test setup, the interior boundary condition was set as constant at 21 °C and 50% RH.

3.4. Parameter Study

As described in previous chapters, hygrothermal simulations were performed using two different test setups, two different RH scenarios at the exterior side, and five different types of concrete. Simulations were also performed with and without water-repellent surface treatment on the exterior surface. An overview of the different variants is provided in Table 3.

Table 3. Simulation variants.

Types of Concrete	Current Test Setup				Modified Test Setup			
	Water-Repellent Surface Treatment		No Surface Treatment		Water-Repellent Surface Treatment		No Surface Treatment	
	"Worst Case"	"High"	"Worst Case"	"High"	"Worst Case"	"High"	"Worst Case"	"High"
Old *	√	√	√	√	√	√	√	√
C12/15	√	√	√	√	√	√	√	√
MASEA	√	√	√	√	√	√	√	√
C35/45	√	√	√	√	√	√	√	√
Waterproof	√	√	√	√	√	√	√	√

* Liquid transfer coefficient is not included for this concrete.

Additional simulations were performed for:

- an alternative plywood board (Plywood USA);
- an alternative generic air layer (air layer of 100 mm without additional moisture capacity);
- rain loads lower than 15 L/m^2.h (i.e., 1.5, 0.11, and 0.01);
- adhering fractions of rain between 0 and 1.

4. Results

Hygrothermal simulations were performed for a façade system subjected to accelerated ageing in the climate simulator. The results of the simulations of the current test setup show that the variation of the moisture content in the PIR insulation varied significantly depending on the type of concrete (see Figure 4 and Table 4). Minor differences occurred when comparing the two RH-scenarios "high" and "worst case". The water-repellent surface treatment in the simulations (including or excluding the liquid uptake of rain) affected the results differently for the various types of concrete. Two types of concrete achieved a low moisture content in the PIR insulation when surface treatment was applied and higher moisture content when rain was allowed to penetrate the surface. Two types of concrete resulted in high moisture content even though surface treatment was applied and somewhat lower moisture content when rain was included. The modified test setup resulted in higher moisture contents in the PIR-insulation than the current test setup, in all the simulated cases.

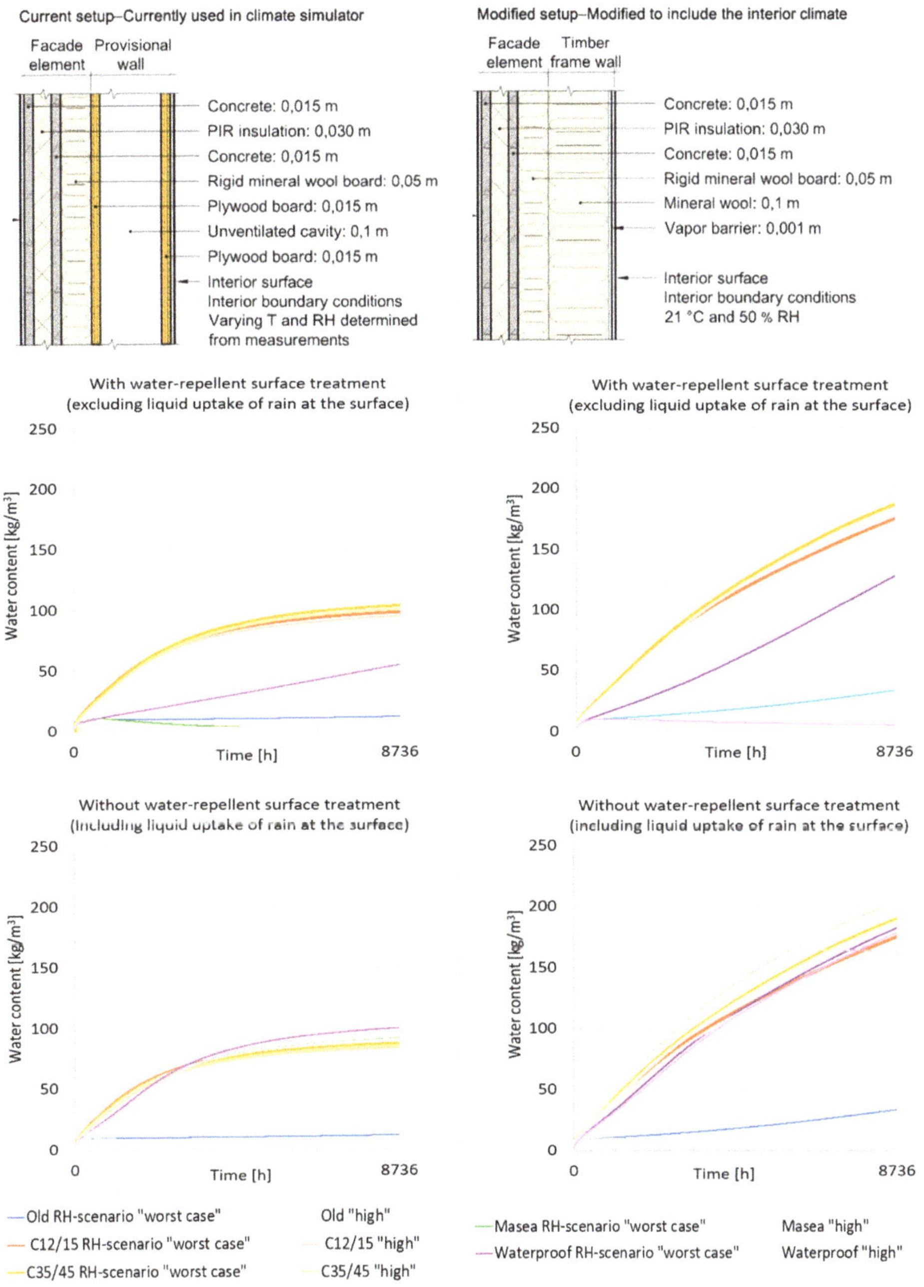

Figure 4. Moisture content in the PIR insulation of the façade solution subjected to accelerated ageing in climate simulator. The left and the right diagrams show the current and the modified test setups, respectively. The upper and lower diagrams show results with and without a water-repellent surface treatment, respectively. The five types of concrete are depicted in different colours. The light and dark colours distinguish between the two RH-scenarios.

Table 4. Moisture content in PIR insulation after one year of accelerated ageing.

Type of Concrete	Current Test Setup				Modified Test Setup			
	Water-Repellent Surface Treatment		No Surface Treatment		Water-Repellent Surface Treatment		No Surface Treatment	
	"Worst Case"	"High"	"Worst Case"	"High"	"Worst Case"	"High"	"Worst Case"	"High"
Old *	12.6	6.9	12.6	6.9	34.0	23.0	33.6	23.0
C12/15	99.0	95.2	87.3	84.3	176.1	171.4	176.1	171.4
MASEA	3.0	2.1	91.6	91.1	5.12	2.8	203.1	202.8
C35/45	104.3	100.8	88.2	85.3	187.9	183.4	190.8	187.4
Waterproof	55.29	3.5	100.4	97.4	128.2	5.1	182.4	178.6

* Liquid transfer coefficients are not included for this concrete; results are thus equal both with and without surface treatment.

5. Discussion

This study investigates the risk of moisture accumulation in a multi-layered façade system exposed to accelerated ageing in the climate simulator according to NT Build 495 [8], and considers the need to change the current test setup by controlling the interior climate. The results and answers to the research questions are discussed in the following section.

5.1. Moisture Accumulation in a Façade System Exposed to Accelerated Ageing

The hygrothermal simulations show that moisture may accumulate in the thermal insulation; however, this is strongly dependent on the concrete quality and whether a water-repellent surface treatment is applied. The C12/15 and C35/45 concretes have the greatest ability to transfer moisture by capillary action (large liquid transfer coefficients) and result in the largest moisture accumulation. The typical application of these concretes is not stated in the WUFI®Pro material database, and the correspondence to the concrete in the façade system is unknown. However, the high moisture uptake suggests that they will not work well in a façade with single-stage protection.

The results also show, in line with previous research [30], that applying a water-repellent surface treatment contributes to prevent moisture accumulation in the façade system. When rain is allowed to penetrate the surface (no water-repellent surface treatment is applied), all relevant concrete characteristics result in high moisture content in the PIR insulation. When a surface treatment is applied (liquid uptake of rain is excluded), the results vary more. The MASEA and Old concretes results in a low moisture content, the waterproof concrete has a low or a moderate moisture accumulation depending on the RH-scenario, and the C12/15 and C35/45 concretes results in high moisture content.

Comparing the results for the C12/15 and C35/45 concrete qualities with and without a surface treatment, is interesting. When rain is allowed to penetrate the surface, the concretes result in lower moisture accumulation in the PIR insulation than when rain is excluded. This behaviour is caused by the mathematical model and the procedure for calculating the capillary absorption of rain. In short, when rain is applied to the component for capillary absorption, the amount of water absorbed during the time step depends on the capillary properties of the surface material, the current saturation state of the material, and the amount of rain available for absorption. Rain is also used to determine whether capillary absorption or capillary redistribution processes are currently dominant in the building component. During the time steps rain is applied, the liquid transport coefficients for suction are used to compute all the capillary transport processes in the component. Otherwise, the liquid transport coefficients for redistribution are used. Due to the fact that the surfaces of the two types of concrete are already close to saturation when rain is applied for adsorption, most of the rain runs off instead of being absorbed. The rate of the moisture transfer in the other capillary material layers in the component, however, is increased. As a result, more moisture is transferred from the PIR insulation towards the interior side during rain events, resulting in a somewhat lower final moisture content in the PIR insulation despite the lack of a water-repellent surface treatment. It is uncertain whether the qualities of these concretes are suitable for the exterior surface of façades with only single-stage protection because of the risk of degradation by freezing and thawing.

5.2. Controlling the Interior Climate

Artificial accelerated ageing of building materials and products are performed according to NT Build 495 [8]. The test is applied to façade systems to assess the durability of their exterior surfaces as withstanding UV light, heat, water, and frost and are crucial to a buildings' climate shell/weather skin. In the current test setup, façade systems are mounted on a provisional wall and subjected to degrading climatic factors on the exterior side. One concern with the current test setup is that the façade systems are tested without the interior parts of the wall assemblies and without a habitable indoor climate. The concern is that that this setup might negatively affect the moisture performance. However, the simulations show that the current test setup results in lesser moisture accumulation in the thermal insulation compared to that of the modified test for all the investigated types of concrete. This is mainly because the interior temperature in the current test setup follows variations in the exterior surface temperature more closely than in the modified setup where the interior parts of the wall is included and the indoor climate is controlled at 21 °C. The interior parts of the wall also limit the drying to the interior side to a certain extent. According to Daniotti et al. [3], more water absorption detriments both ageing and thermal performance of ETICS.

5.3. Limitations of The Study

In this study, we assess the possible moisture accumulation in the PIR insulation in a façade solution that may occur without damage/degradation at the exterior surface. It was therefore assumed that the façade system was subjected to accelerated ageing for 12 months without cracking of the exterior surface and without significant changes to the hygrothermal material properties. To gain information on the durability, testing by either natural ageing or accelerated artificial climatic ageing must be carried out.

Modelling the complex coupled heat and moisture transfer processes in building components always involves simplification of reality. Some materials (e.g., concrete) do not conform to the simplified transport equations and change their material data depending on their present and past moisture content. Materials with a pronounced hysteresis in their moisture storage function (e.g., concrete) may not be sufficiently described using an averaged moisture-storage function. The total moisture transfer resulting from the combination of liquid and vapour transport processes under varying thermal conditions, as in the climate simulator, is also difficult to calculate because the two flows cannot be divided in laboratory experiments. The errors caused by these general inaccuracies may be negligible or serious. To determine the reliability of the calculations, the results should be compared with measurements.

The boundary conditions applied in the hygrothermal simulations were determined based on measurements of temperature and RH in the climate simulator. Due to the fact that the RH humidity at the exterior surface (exterior boundary condition) was lacking, two different RH scenarios were determined from measurements of RH in the air gap behind the test samples: "Worst case" and "High". The two determined scenarios were considered conservative; that is, higher than what the measurements indicated, in particular during exposure to the ambient laboratory climate. The two conservative scenarios were considered appropriate for the scope of this study; however, the moisture accumulation may be somewhat overestimated in this study because of this inaccuracy. If the purpose of the simulations had been to determine the amount of moisture accumulation more precisely, more detailed measurements of the boundary conditions and material properties would have been necessary.

The material properties of the products in the façade system in this study have not been documented. Ideally, the material properties should have been determined experimentally under similar boundary conditions as in practical use. This was beyond the scope of this study. Therefore, five different types of concrete derived from the WUFI®Pro material database were compared. The correspondence of these concretes to the actual material properties of the concrete in the façade system is unknown.

As discussed in this chapter, there were inaccuracies in the calculations caused by the limitations in the hygrothermal model, the determination of boundary conditions, and the lack of documented material properties. A main concern is the unknown material properties of the concrete and the modelling of rain absorption at the surface. The ability of the simulation software to realistically replicate the moisture transfer in multi-layered façade systems exposed to rain events over long time periods is questionable. Although these inaccuracies must be acknowledged when interpreting the calculated moisture accumulation, they are considered to be of less importance for the main results and conclusions in this article.

6. Conclusions

The hygrothermal simulations of a façade system exposed to accelerated ageing testing according to NT Build 495 [8], show that the amount of moisture accumulation depends strongly on the type of concrete and whether a water-repellent surface treatment is applied. The results show that a water-repellent surface treatment contributes to preventing moisture accumulation in the façade system. The two types of concrete with the greatest ability to absorb moisture by capillary forces resulted in the highest moisture content in the thermal insulation at the end of the test. It is however uncertain whether these two types of concrete are suitable for the façade system with only single-stage protection because of the high moisture uptake at the exterior surface. The material properties of the products in the façade system in this study have not been determined through testing. To calculate the moisture accumulation accurately, the material properties must be determined experimentally under similar boundary conditions as in practical use.

The current test setup resulted in lower moisture accumulation compared to that of the modified test setup for all the investigated types of concrete. More water absorption may lead to accelerated degradation and negatively affect the thermal performance.

Author Contributions: Conceptualization, S.K.A., B.T. and T.K.; methodology, S.K.A., B.T. and T.K.; formal analysis, S.K.A., B.T. and T.K.; investigation, S.K.A.; data curation, S.K.A.; writing—original draft preparation, S.K.A.; writing—review and editing, S.K.A., B.T. and T.K.; visualization, S.K.A.; supervision, B.T. and T.K.; funding acquisition, B.T. and T.K. All authors have read and agreed to the published version of the manuscript.

Funding: This research and the APC were funded by SINTEF, The Research Council of Norway, and several partners through the Centre for Research-based Innovation 'Klima 2050', grant number 237859.

Institutional Review Board Statement: Not applicable.

Informed Consent Statement: Not applicable.

Acknowledgments: The authors would like to extend a special thanks to Sivert Uvsløkk and Egil Rognvik for conducting and assessing the measurements of temperature and RH performed in the climate simulator, and to CAD operator Anne Eliassen for drawing Figure 2.

Conflicts of Interest: The authors declare no conflict of interest.

Appendix A

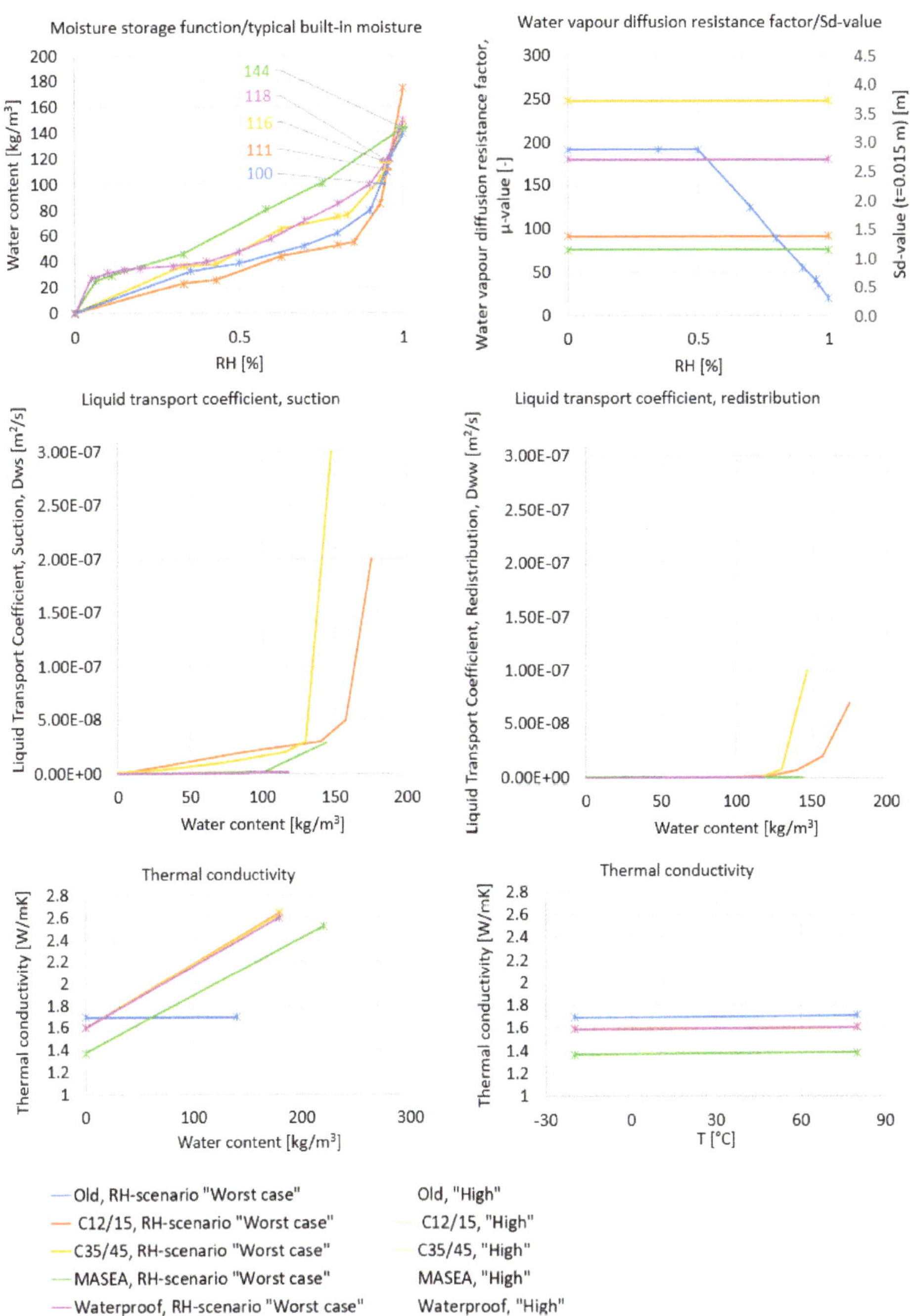

Figure A1. Material properties of the five types of concrete used in the simulations. The graphs are created by the authors based on selected material data from the WUFI®Pro material database [17].

Table A1. Properties of materials used in simulations. The material properties are mainly obtained from the WUFI®Pro material database [17]. The exception is the material properties of the PIR insulation, which is defined based on input from the producer of the façade solution. The images illustrating the material properties are obtained from the WUFI®Pro Software [17].

Material	PIR	Mineral Wool Board	Plywood	Plywood	Mineral Wool	Vapour Barrier	Air Cavity
Name of material in the WUFI®Pro material database	User defined *	Isover Ultimate Klemmfilz–035 (Fraunhofer-IBP)	Plywood low (North America database)	Plywood USA (North America database)	Mineral wool (heat cond.: 0.04 W/mK) (Fraunhofer-IBP)	Vapour retarder s_d = 10 m (Fraunhofer-IBP)	Air layer 100 mm ** (generic materials)
Thickness (m)	0.03	0.05	0.015	0.015	0.1	- (0.001)	0.1
s_d-value (m)	1.38	0.05	RH0%: 7.4, 00%: 0.14	RH0%: 16.3, 100%: 0.33	0.13	10	0.015
Water vapour diffusion resistance factor, μ (-), RH dependent	46	1	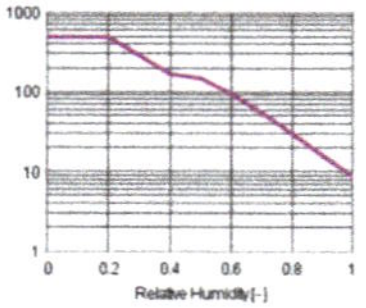 493-9	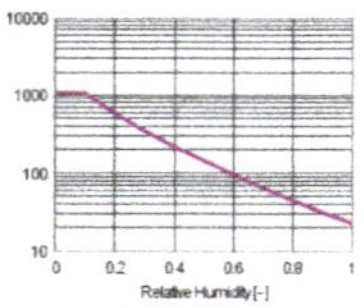 1088-22	1.3	10,000	0.15
Moisture storage function (kg/m^3) RH dependent	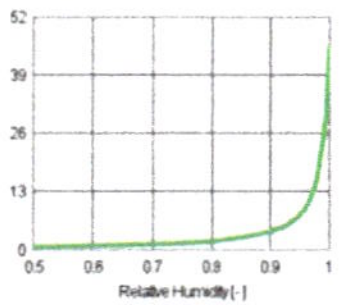	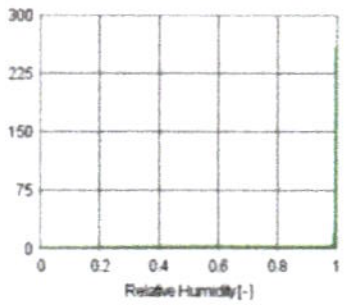	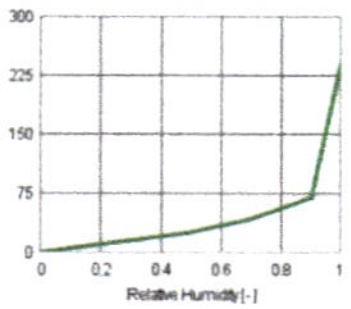	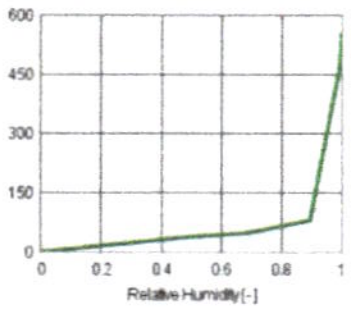	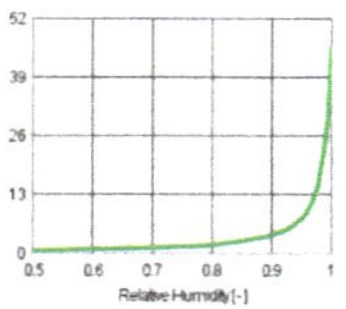	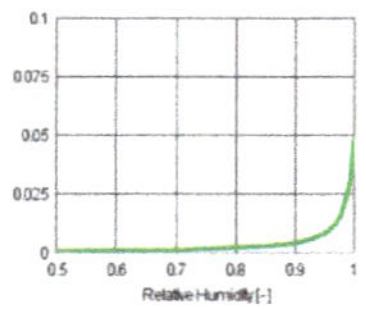	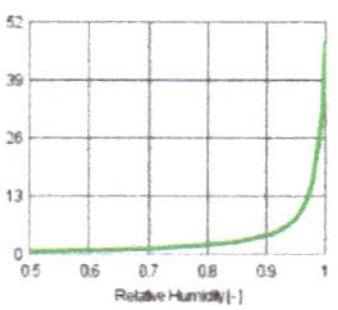
Liquid transport coefficient, suction, Dws (m^2/s) dependent on moisture content (kg/m^3)	***	***	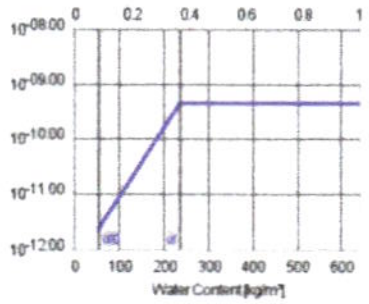	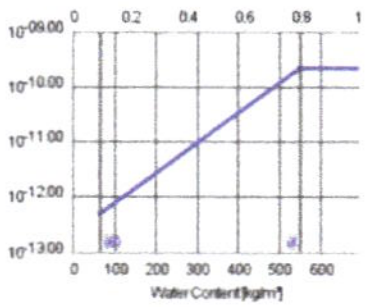	***	***	***

Table A1. *Cont.*

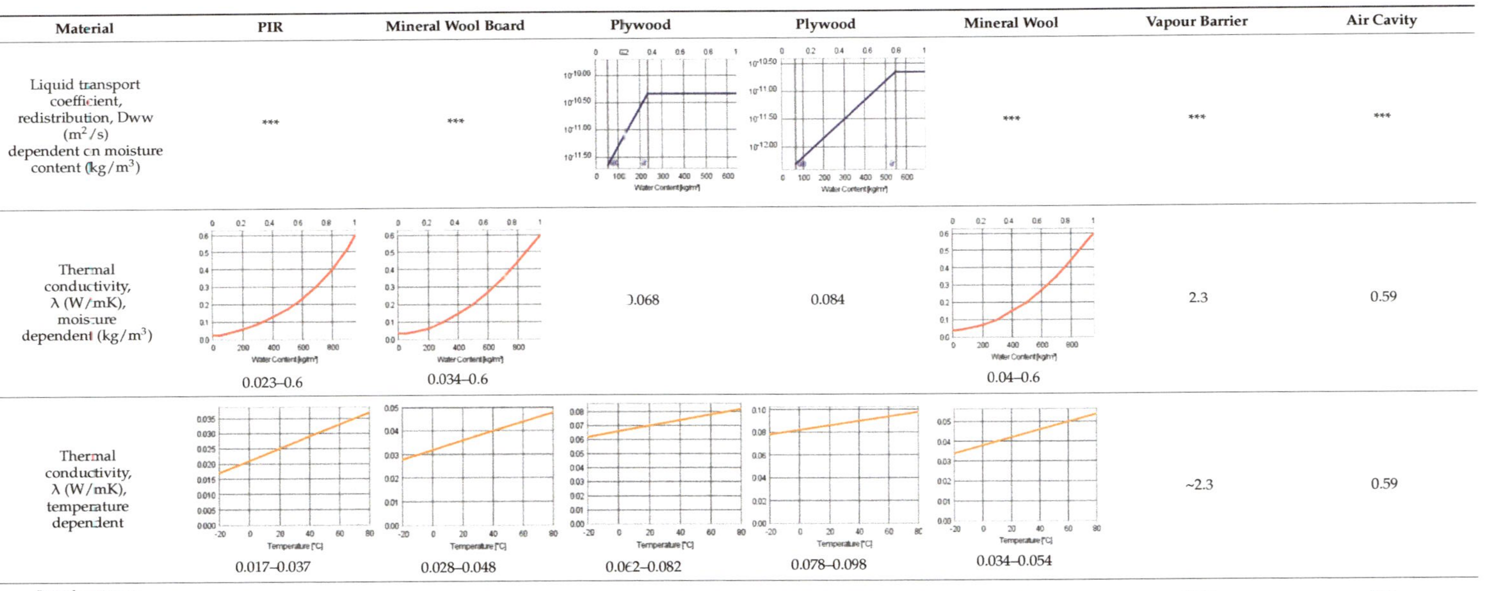

Material	PIR	Mineral Wool Board	Plywood	Plywood	Mineral Wool	Vapour Barrier	Air Cavity
Liquid transport coefficient, redistribution, Dww (m²/s) dependent on moisture content (kg/m³)	***	***	(graph)	(graph)	***	***	***
Thermal conductivity, λ (W/mK), moisture dependent (kg/m³)	(graph) 0.023–0.6	(graph) 0.034–0.6	0.068	0.084	(graph) 0.04–0.6	2.3	0.59
Thermal conductivity, λ (W/mK), temperature dependent	(graph) 0.017–0.037	(graph) 0.028–0.048	(graph) 0.062–0.082	(graph) 0.078–0.098	(graph) 0.034–0.054	~2.3	0.59
Initial moisture (kg/m³) ****	1.8	0.5	55	70	1.8	0.002	0.01

* The material properties of the PIR insulation are defined based on input from the producer of the façade solution. The s_d-value, μ-value, and thermal conductivity were specified. The moisture storage function and initial moisture content are based on properties provided by the WUFI®Pro material database. ** According to WUFI®, generic air layers can be used for unventilated air cavities between non-metallic surfaces. The effective material parameters of the air layers describe the total heat transport (owing to heat conduction, free convection, and radiative transport), and total vapour transport (owing to diffusion and free convection). For each layer thickness, the database provides two air layers with different moisture storage functions: a standard air layer and an air layer without additional moisture capacity. The latter has a lower (more realistic) moisture storage function which corresponds to the moisture contents of a 20 °C warm air layer. The moisture contents and hygric inertia of these air layers are at a realistic level; however, they may prove numerically challenging. The standard layers use the program's default moisture storage function, resulting in an unrealistically high moisture content for an air layer. These moisture contents are less demanding on the numerics, but should not be used for evaluating the construction. *** Not included in WUFI®Pro material database since the respective materials are considered not to be capillary active. **** The typical built-in moisture contents suggested by WUFI®Pro were used as initial conditions.

Appendix B

Appendix B.1. Measurements

Temperature and RH measurements were performed on two test elements subjected to accelerated artificial ageing in SINTEF's climate simulator. The two test elements consisted of parts of a façade system previously subjected to accelerated ageing according to NT Build 495 [8]. The two test elements were of the same build as the façade solution investigated in the simulations but without the mineral wool board and the provisional wall. The measured data were considered representative for the façade solution investigated in the simulations in this study.

Temperatures were measured on the exterior surfaces and in the air gap behind the test elements. The RH was measured in the air gap behind the two test elements. The sensors on the outside (outdoor climate) were positioned on the surface of the test elements, and the sensors on the back of the test elements (indoor climate) were positioned in the air gap at the back. The RH on the surface of the test element used in the simulation was partly calculated and partly determined by assessment as described below.

Based on the measured data, representative climate data for 4-h were constructed which corresponded to an entire cycle in the climate simulator. In the simulations, these four hours were repeated for one year (see Figure 2 in the main text). The temperature and RH were measured with minute values because the changes in the climate simulator during a 4-h period were expected to vary continuously. In WUFI®Pro, hourly values are commonly used. In these simulations, 0.1-h (6 min) was considered to be a suitable time step (see Figure 2 in the main text).

The exterior boundary conditions (applied in the simulations of both the current and the modified test setups) are determined based on temperatures and the RH measured at the surface of the test element. The temperatures used were the same as those measured in the middle of the surface of one of the test elements. The selected measurement period was 20 March 2020, 15.00 to 18.54. The measured values were converted from minute values to 6-min values. The software WUFI®Pro calculates the surface temperature of a sample if the air temperature and radiation data are provided. In this case, the measurements of the surface temperature were used directly to account for solar radiation.

As there were no measurements of RH on the exterior surface of the test elements, the RH on the surface (exterior boundary condition) of the façade solution in the simulation was partly calculated and partly determined by assessment. Two conservative climate scenarios were determined and used in the simulations: "high" and "worst case". Both climate scenarios represented RH variations which were probably higher than what would have been measured. By making further assessments/assumptions, a third climate scenario with a lower RH can be determined. A lower RH would result in a lower moisture content in the PIR-insulation. Due to the fact that this will only have a positive effect and does not affect the main results in the report, these simulations were not performed. The RH on the surface of the façade element was first calculated from measurements of RH and temperatures in the middle of the air space at the back of the test element along with the temperatures measured on the surface of one of the test elements. The RH was calculated using equations for the water vapour saturation pressure as a function of temperature. The equations given in the WUFI®Pro online help were used [17,29]. In the rain chamber, the RH at the front of the test element was considered to be 100%. There was more uncertainty associated with the RH on the front of the test element when it was in the freezing chamber or in the laboratory room. In the freezing chamber, it is conceivable that a water film froze on the surface of the test element. The RH will then be close to 100% which is far higher than the calculated RH based on measurements in the air behind the test element. In the laboratory room, the RH in the airspace behind the test element was 100%. The calculated RH on the front will then be over 100% and can therefore not be used. However, measurement of RH and temperature in the laboratory near the climate simulator for the relevant measurement period showed a low RH (3–8% at 20 °C). The calculated RH for the test element surface based on the vapour pressure measured during the period when the test element was in the laboratory decreased from 100% to 7% RH in 1 h. The correct RH was considered to be somewhere between these extremes. Due to the fact that there was some uncertainty associated with the RH on the exterior surface, two different climate scenarios were used in the simulations in this study (see Table 2 and Figure 3 in the main text).

Appendix B.2. Rain

In the climate simulator, tested façade systems are subjected to a water spray for 50 min each quarter hour. The amount of rain in the simulation was therefore set at 15 L/m^2.h, which corresponds to the 15 mm of rain per hour of the climate simulator. The proportion of rain that hit the surface of the façade solution ("adhering fraction of rain") was set to 1. This means that all the rain is made available to the surface. The proportion of rain drawn into the material surface in the simulation then is dependent on the surface layer and the moisture transfer properties of the concrete.

Appendix B.3. Surface Treatment on Exterior Surface

In the simulations, two different situations were studied (see Table 4). First it was assumed that the surfaces of the façade solution were treated with a water-repellent diffusion open-surface treatment; in this case the water hitting the exterior surface in the rain chamber was not drawn into the surface capillary. Then, it was assumed that the surface of the façade solution was untreated and thus that the water that hit the exterior surface in the rain chamber was drawn into the surface depending on the properties of the concrete.

Appendix B.4. Boundary Conditions on Interior Side

In simulations of the current test setup, the measured temperature and RH in the air gap at the back of the test element were used for the interior surface of the façade solution. The selected measurement period was 20 March 2020, 15.00 to 18.54. The measured values were converted from minute values to 6-min values. A surface transfer resistance of 0.13 m^2 K/W was applied to the interior surface.

In simulations of the modified test setup, the provisional wall was replaced with 100-mm mineral wool and a vapour barrier, and a constant climate of 21 °C and 50% RH were used for the interior surface of the façade solution. A surface transfer resistance of 0.13 m^{2}K/W was applied to the interior surface.

References

1. Jelle, B.P. Accelerated climate ageing of building materials, components and structures in the laboratory. *J. Mater. Sci.* **2012**, *47*, 6475–6496. [CrossRef]
2. Hanssen-Bauer, I.; Førland, E.J.; Haddeland, I.; Hisdal, H.; Mayer, S.; Nesje, A.; Nilsen, J.E.Ø.; Sandven, S.; Sandø, A.B.; Sortberg, A.; et al. *Climate in Norway 2100 a Knowledge Base for Climate Adaptation*; The Norwegian Centre for Climate Services (NCCS): Bergen, Norway, 2017.
3. Daniotti, B.; Paolini, R.; Re Cecconi, F.R. Effects of Ageing and Moisture on Thermal Performance of ETICS Cladding. In *Durability of Building Materials and Components*; Building Pathology and Rehabilitation; de Freitas, V.P., de Delgado, J.M.P.Q., Eds.; Springer: Berlin, Heidelberg, 2013; pp. 127–171.
4. Ibrahim, M.; Wurtz, E.; Biwole, P.; Achard, P.; Sallee, H. Hygrothermal performance of exterior walls covered with aerogel-based insulating rendering. *Energy Build.* **2014**, *84*, 241–251. [CrossRef]
5. Ihara, T.; Jelle, B.P.; Gao, T.; Gustavsen, A. Aerogel granule aging driven by moisture and solar radiation. *Energy Build.* **2015**, *103*, 238–248. [CrossRef]
6. Kvande, T.; Bakken, N.; Bergheim, E.; Thue, J.V. Durability of ETICS with Rendering in Norway—Experimental and Field Investigations. *Buildings* **2018**, *8*, 93. [CrossRef]
7. Jelle, B.P.; Nilsen, T.-N.; Hovde, P.J.; Gustavsen, A. Accelerated climate aging of building materials and their characterization by Fourier transform infrared radiation analysis. *J. Build. Phys.* **2011**, *36*, 99–112. [CrossRef]
8. NT Build 495. *Building Materials and Components in the Vertical Position: Exposure to Accelerated Climatic Strains*; Nordtest: Espoo, Finland, 2000.

9. Francke, B.; Zamorowska, R. Resistance of External Thermal Insulation Composite Systems with Rendering (ETICS) to Hail. *Materials* **2020**, *13*, 2452. [CrossRef]
10. Asphaug, S.K.; Jelle, B.P.; Gullbrekken, L.; Uvsløkk, S. Accelerated ageing and durability of double-glazed sealed insulating window panes and impact on heating demand in buildings. *Energy Build.* **2016**, *116*, 395–402. [CrossRef]
11. Fufa, S.M.; Labonnote, N.; Frank, S.; Rüther, P.; Jelle, B.P. Durability evaluation of adhesive tapes for building applications. *Constr. Build. Mater.* **2018**, *161*, 528–538. [CrossRef]
12. Rüther, P.; Jelle, B.P. Color changes of wood and wood-based materials due to natural and artificial weathering. *Wood Mater. Sci. Eng.* **2013**, *8*, 13–25. [CrossRef]
13. Wegger, E.; Jelle, B.P.; Sveipe, E.; Grynning, S.; Gustavsen, A.; Baetens, R.; Thue, J.V. Aging effects on thermal properties and service life of vacuum insulation panels. *J. Build. Phys.* **2011**, *35*, 128–167. [CrossRef]
14. Gonçalves, M.; Simões, N.; Serra, C.; Flores-Colen, I. Laboratory assessment of the hygrothermal performance of an external vacuum-insulation composite system. *Energy Build.* **2021**, *254*, 111549. [CrossRef]
15. Hens, H.S.L.C.; Mukhopadhyaya, P.; Kumaran, M.; Dean, S.W. Modeling the Heat, Air, and Moisture Response of Building Envelopes: What Material Properties are Needed, How Trustful Are the Predictions? *J. ASTM Int.* **2007**, *4*, 460. [CrossRef]
16. EOTA European Assessment Document EAD 040083-00-0404. External Thermal Insulation Composite Systems (ETICS) with Renderings. 2019. Available online: https://www.eota.eu/download?file=/2014/14-04-0083/for%20ojeu/ead%20040083-00-0404_ojeu2020.pdf (accessed on 1 November 2021).
17. Fraunhofer IBP. WUFI, WUFI®Pro 6.5. 2021. Available online: https://wufi.de/en/ (accessed on 17 August 2021).
18. Kvande, T.; Lisø, K.R. Climate adapted design of masonry structures. *Build. Environ.* **2009**, *44*, 2442–2450. [CrossRef]
19. Passa, D.S.; Sotiropoulou, A.B.; Pandermarakis, Z.G.; Mitsopoulos, G.D. Thermal and Drying Cyclic Loading for Cement Based Mortars and Expanded Polystyrene Foam Layers. *Appl. Mech. Mater.* **2012**, *204–208*, 3648–3651. [CrossRef]
20. Lisø, K.R.; Kvande, T.; Thue, J.V. High-performance weather-protective flashings. *Build. Res. Inf.* **2005**, *33*, 41–54. [CrossRef]
21. De Freitas, S.S.; de Freitas, V.P. Cracks on ETICS along thermal insulation joints: Case study and a pathology catalogue. *Struct. Surv.* **2016**, *34*, 57–72. [CrossRef]
22. Griciutė, G.; Bliūdžius, R.; Norvaišienė, R. The Durability Test Method for External Thermal Insulation Composite System used in Cold and Wet Climate Countries. *J. Sustain. Arch. Civ. Eng.* **2013**, *1*, 50–56. [CrossRef]
23. D'Orazio, M.; Stipa, P.; Sabbatini, S.; Maracchini, G. Experimental investigation on the durability of a novel lightweight prefabricated reinforced-EPS based construction system. *Constr. Build. Mater.* **2020**, *252*, 119134. [CrossRef]
24. Franzoni, E.; Pigino, B.; Graziani, G.; Lucchese, C.; Fregni, A. A new prefabricated external thermal insulation composite board with ceramic finishing for buildings retrofitting. *Mater. Struct.* **2015**, *49*, 1527–1542. [CrossRef]
25. Parracha, J.L.; Borsoi, G.; Veiga, R.; Flores-Colen, I.; Nunes, L.; Garcia, A.R.; Ilharco, L.M.; Dionísio, A.; Faria, P. Effects of hygrothermal, UV and SO_2 accelerated ageing on the durability of ETICS in urban environments. *Build. Environ.* **2021**, *204*, 108151. [CrossRef]
26. Pereira, M.C.; Soares, A.; Flores-Colen, I.; Correia, J.R. Influence of Exposure to Elevated Temperatures on the Physical and Mechanical Properties of Cementitious Thermal Mortars. *Appl. Sci.* **2020**, *10*, 2200. [CrossRef]
27. Maia, J.; Pedroso, M.; Ramos, N.; Pereira, P.; Flores-Colen, I.; Gomes, M.G.; Silva, L. Hygrothermal performance of a new thermal aerogel-based render under distinct climatic conditions. *Energy Build.* **2021**, *243*, 111001. [CrossRef]
28. Maia, J.; Ramos, N.M.M.; Veiga, R. Assessment of test methods for the durability of thermal mortars exposure to freezing. *Mater. Struct.* **2019**, *52*, 112. [CrossRef]
29. Künzel, H.M. *Simultaneous Heat and Moisture Transport in Building Components*; One- and Two-Dimentional Calculation Using Simple Parameters; IRB-Verlag: Stuttgart, Germany, 1995; ISBN 3-8167-4103-7.
30. Slapø, F.; Kvande, T.; Bakken, N.; Haugen, M.; Lohne, J. Masonry's Resistance to Driving Rain: Mortar Water Content and Impregnation. *Buildings* **2017**, *7*, 70. [CrossRef]

Article

Improvement of the Inspection Interval of Highway Bridges through Predictive Models of Deterioration

Ademir F. Santos [1], Maurício S. Bonatte [1,*], Hélder S. Sousa [1], Túlio N. Bittencourt [2] and José C. Matos [1]

1 Department of Civil Engineering, Institute for Sustainability and Innovation in Structural Engineering (ISISE), University of Minho, 4800-058 Guimarães, Portugal; adsantos28@gmail.com (A.F.S.); sousa.hms@gmail.com (H.S.S.); jmatos@civil.uminho.pt (J.C.M.)
2 Department of Structural and Geotechnical Engineering, University of São Paulo, São Paulo 05508-900, Brazil; tbitten@usp.br
* Correspondence: mbonatte@ymail.com

Abstract: Bridges have substantial significance within the transport system, considering that their functionality is essential for countries' social and economic development. Accordingly, a superior level of safety and serviceability must be reached to ensure the operating status of the bridge network. On that account, the recent collapses of road bridges have led the technical–scientific community and society to reflect on the effectiveness of their management. Bridges in a network are likely to share coinciding environmental conditions but may be subjected to distinct structural deterioration processes over time depending on their age, location, structural type, and other aspects. This variation is usually not considered in the bridge management predictions. For instance, the Brazilian standards consider a constant inspection periodicity, regardless of the bridges' singularities. Consequently, it is helpful to pinpoint and split the bridge network into classes sharing equivalent deterioration trends to obtain a more precise prediction and improve the frequency of inspections. This work presents a representative database of the Brazilian bridge network, including the most relevant data obtained from inspections. The database was used to calibrate two independent predictive models (Markov and artificial neural network). The calibrated model was employed to simulate different scenarios, resulting in significant insights to improve the inspection periodicity. As a result, the bridge's location accounting for the differentiation of exposure was a critical point when analyzing the bridge deterioration process. Finally, the degradation models developed following the proposed procedure deliver a more reliable forecast when compared to a single degradation model without parameter analysis. These more reliable models may assist the decision process of the bridge management system (BMS).

Keywords: highway bridges; bridge inspection; predictive models; Markov; ANN

Citation: Santos, A.F.; Bonatte, M.S.; Sousa, H.S.; Bittencourt, T.N.; Matos, J.C. Improvement of the Inspection Interval of Highway Bridges through Predictive Models of Deterioration. *Buildings* **2022**, *12*, 124. https://doi.org/10.3390/buildings12020124

Academic Editors: Ana Silva and Vitor Silva

Received: 15 December 2021
Accepted: 21 January 2022
Published: 26 January 2022

Publisher's Note: MDPI stays neutral with regard to jurisdictional claims in published maps and institutional affiliations.

1. Introduction

A successful Bridge management system (BMS) depends heavily on defining appropriate intervention actions to ensure structural safety, functionality, and durability while maintaining the lowest financial investment related to the available budget [1]. By accounting for an adequate quality control plan and a risk classification, a prolonged quality assurance of bridges can be assured and, consequently, a proper allocation of funds [2–4].

One of the keys to successful asset management is the use of predictive models that allow foreseeing, for different periods, the performance of the asset, taking into account the demand values of exposure. Thus, the subsequent modules related to the time and extent of the necessary maintenance actions depend entirely on the established deterioration model, the consequences triggered in case of failure, and the costs of each type of intervention [5].

Many investigations have attempted to improve the deterioration modelling. Ref. [1] set a probabilistic model (two-dimensional Markov process) to predict bridge deterioration and define the optimal inspection intervals. Ref. [6] applied a two-step cluster analysis to identify the most critical parameters, such as age, distance from coastline, and climatic regions.

Numerous researchers have endeavoured to enhance the deterioration modelling to predict the remaining service life of bridges in Brazil. Ref. [7] presents the results of deterioration rates of constructions in 16 studied road segments, based on a methodology that uses the Markov chains method. During the study, numerous reports of inspections on bridges were collected, summing up 1707 bridges inspected in an average of 7 years. In addition to verifying the rates of deterioration and their relationship with possible agents of degradation, ref. [7] also contributes to the knowledge of inspection practices carried out in Brazil and their effectiveness in its use for the administration of national road bridges. On the other hand, ref. [8] analysed the pathological manifestations and the structural deficiencies of bridges and viaducts of the federal highways in Pernambuco (a Brazilian state). The obtained results allowed the authors to present the current situation of the investigated bridges. The work aims to subsidise responsible public agencies' decision-making, planning maintenance, thus ensuring more outstanding durability and valuable life for the bridges.

Additionally, ref. [9] presents the status of the bridges on Brazilian federal highways, based on data obtained from National Department of Transport Infrastructure (DNIT), Institute of Road Research (IPR), and National Land Transport Agency (ANTT), among others, which constitute a register with 5619 bridges, with levels of information that vary in dimensions, inspection results, sketches, photos, and geographic coordinates. The analysis of these data provided more significant knowledge about the reality of bridges on Brazilian federal highways, producing subsidies for the planning of a bridge management system that is more compatible with reality and led to an understanding of the main aspects that guide state assessments of the bridges. According to [10], the reinforced concrete beam is the most used system (2764 bridges), representing more than 58%, followed by a reinforced concrete slab (777 bridges) and the prestressed concrete (622 bridges), within a spectrum of 4725 bridges. Almost 25.7% of the bridges in the inventory were constructed before 1960, and 52% were built between 1960 and 1975.

Brazil has an expressive set of bridges, with about 120,000 bridges [11] distributed sparsely around the entire country, exposed to different environmental conditions. Although most of these bridges have not been catalogued, this article gathers a comprehensive database containing 10,331 bridges. The database includes geometric information, design parameters, operating circumstances, and structural conditions. The inventory has specific limitations regarding the inspections records, such as a short time window of inspection for many bridges and large amounts of data scatter, added to the subjectivity of the visual inspection itself that is also a limitation.

Currently, three standards establish the conditions required to carry out inspections and present the results in Brazil. The DNIT-010 standard [12] has been used to evaluate bridges located on highways under the jurisdiction of the Federal Government. NBR 9452 standard [13] proposes to assess structural safety in a similar way to DNIT-010 standard [12], adding indicators related to durability and functionality. The standard published by the São Paulo State Transport Agency (ARTESP [14]) is responsible for regulating state highways granted in the State of São Paulo, and it adds the concept of "urgent intervention".

The present study aims at applying two probabilistic models, Markov and Artificial neural network (ANN), to forecast bridge deterioration and improve the inspection periodicity proposed by the standards [12–14]. The methodology proposed in this study is implemented using a dataset encompassing information about inspections of 10,331 bridges throughout Brazil from 2008 to 2021. The investigation results can assist bridge owners and transport agencies in efficiently allocating maintenance resources and invest the capital currently allocated for unnecessary inspections in desired infrastructure development projects.

Section 2 discusses the current inspection standards, their applicability and limitations, focusing on the periodicity of inspections, followed by Section 3, which presents a state-of-art of the two predictive modeling methods. In Section 4, the most up-to-date bridge inventory is presented with additional statistical discussion. Finally, in Section 5, the methodology and results of the two predictive models are presented.

2. Standards

Given bridge inspection's strategic and economic importance, several governments and research centers are dedicated to standardizing inspection techniques, test methods, and bridge monitoring and management systems. Most of them are linked to government agencies or directly subordinated to the countries' transport departments, along with agreements between governments and universities. Some of the standards on the subject created by leading Brazilian centers and reference researchers in the inspection field of bridges and viaducts will now be detailed.

The DNIT-010 standard [12] has been used to evaluate bridges located on highways under the Federal Government jurisdiction. According to this standard, bridges are classified based on structural safety indexes ranging from 1 to 5, where 1 corresponds to a condition of precarious stability and 5 to an excellent condition of stability, prescribing/specifying inspections every two years.

The NBR 9452 standard [13] proposes assessing structural safety in a similar way to the DNIT-010 standard [12], although it adds indicators related to durability and functionality. The bridges are also classified according to condition indexes ranging from 1 to 5, where 1 corresponds to a critical condition and 5 to an excellent condition. The proposed periodicity of routine inspections is one year, regardless of the class.

The standard published in 2007 by the São Paulo State Transport Agency (ARTESP [14]) is responsible for regulating state roads granted in the State of São Paulo, adding the concept of "urgent intervention". The standard provides a total of eight classes, ranging from C0 (poor condition and urgency of immediate intervention) to A5 (excellent condition and urgency of intervention in 5 years). The periodicity of routine inspections is one year, therefore following the requirements of NBR 9452 [13].

Table 1 presents a qualitative comparison between the various bridge evaluation criteria from the standards. The DNIT [12] and ABNT [13] standards present the same classification criteria related to structural safety (reliability), ranging from 1 to 5. On the other hand, the ARTESP [14] standard classifies the bridges according to eight classification levels of conditions.

Table 1. Qualitative comparison between ABNT/DNIT and ARTESP.

ARTESP	ABNT/DNIT	Condition
A5	5	Excellent condition There is no damage or structural insufficiency Nothing to do
B4/ A4	4	Good condition There is some damage, but there are no signs that they are causing structural failure Nothing to do; maintenance services only
B3/ B2/ C2	3	Apparently good condition There is damage causing some structural insufficiency, but there are no signs of compromised structural stability The recovery of the structure can be postponed. However, in this case, the problem should be placed under systematic observation
C1	2	Poor condition There is damage generating significant structural insufficiency in the bridge, but there is apparently no real risk of structural collapse yet The restoration (usually with structural strengthening) of the bridge must be done in the short term
C0	1	Critical condition/urgent intervention There is damage causing severe structural insufficiency in the bridge; the element is critical, with a real risk of structural collapse Recovery (usually with structural strengthening)—or in some cases, replacement of the bridge—must be done without delay

In the European context, Cost Action TU1406, created in 2015, brings together academic researchers, industrial professionals, European government agencies, and international observers to establish a European guideline on quantifying performance indicators to evaluate the quality control plan. The result of the Cost Action TU1406 was the publication, in 2019, of the quality specifications for roadway bridges, standardization at a European level [4]. The TU1406 proposes five levels to assess the bridge Condition index (CI), ranging from 1 to 5, associated with the urgency of intervention, where 1 corresponds to a bridge in good condition and 5 a bridge in critical condition, requiring immediate intervention. According to [4], the inspections need to be conducted in predefined intervals, but they should rely on bridge condition and bridge significance to the network. Table 2 presents a comparison between the bridge assessment systems presented above.

Table 2. Comparison between standards.

Standard	Year	Performance Indicator						Rate Scale	Frequency of Inspections
		Structural	Safety	Durability	Cost	Environment	Availability		
DNIT-010	2004	✓	✗	✗	✗	✗	✗	5	Biennial
NBR-9452	2019	✓	✓	✓	✗	✗	✗	5	Annual
ARTESP	2007	✓	✓	✓	✗	✗	✗	8	Annual
TU1406	2119	✓	✓	✗	✓	✓	✓	5	Variable

3. Bridge Deterioration Models

Bridge deterioration is the process of decay resulting from normal operating conditions. The deterioration process exhibits the combined physical and chemical transformations occurring in various bridge elements. The situation is moderately complex because each element has its distinctive decay rate. As bridges undergo gradual deterioration processes, they are subject to periodic inspections. The purpose of the inspection is to detect defects that may appear throughout the bridge's life. The records of these inspections can be used to develop bridge deterioration models, allowing to extrapolate the bridge CI over the years. Precisely predicting the deterioration rate of each bridge element is hence vital to the success of any BMS.

Approaches to calculating decay rates for bridge elements can mainly be sorted into three general categories: deterministic method, stochastic approach, and ANN-based model [15]. Deterministic models rely on a mathematical or statistical relationship between the factors that affect bridge degeneration. The outcome of such models is described by deterministic values representing average expected conditions, i.e., there are no probabilities involved. Deterministic models can be carried out by extrapolation, regression, and straight-line curve fitting methods [15]. However, deterministic models neglect the uncertainty inherent in stochastic deterioration nature, are computationally expensive when updating the model, and overlook the interaction between different bridge components [15]. In the following sections, stochastic and ANN models are adequately discussed.

3.1. Stochastic Models

Stochastic processes have been used to model the deterioration of infrastructure over time, such as bridges and roads, due to the random nature inherent to a deterioration process. A widely used stochastic process is the Markov chain process [16–18]. Markov chain models capture the uncertainties and randomness of the deterioration process, accumulating the probability of transition from one condition state to another over several discrete (or continuous) time intervals. The Markov model simplifies the transition probability by defining that the next state only depends on the current state and not on the sequence of preceding ones, as illustrated in Equation (1).

$$P(X_{t+1} = j \mid X_t = i) = P(i, j) \tag{1}$$

The values assumed by i and j are called condition states and are denoted by 5, 4, 3, 2, and 1, where 1 represents the worst CI, according to Table 1.

A Markov deterioration matrix presents the probability that a bridge will shift condition within a specified period, generally considered the time between two central inspections. An example of a Markov deterioration process based on a five condition state model, with 5 as the best CI, is shown in Figure 1.

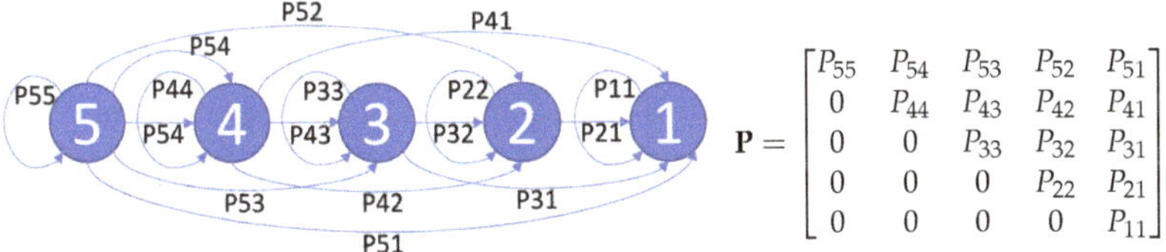

$$\mathbf{P} = \begin{bmatrix} P_{55} & P_{54} & P_{53} & P_{52} & P_{51} \\ 0 & P_{44} & P_{43} & P_{42} & P_{41} \\ 0 & 0 & P_{33} & P_{32} & P_{31} \\ 0 & 0 & 0 & P_{22} & P_{21} \\ 0 & 0 & 0 & 0 & P_{11} \end{bmatrix}$$

Figure 1. Transition probabilities for a 5 state model.

Discrete-time Markov chain is generally used assuming a constant interval between inspections. The implementation of this model simplifies the mathematical formulation and its calculation to obtain the performance prediction curve. However, this assumption does not correspond to reality in many cases since inspections do not occur at uniform intervals.

In the continuous-time Markov chain, the transition between states occurs in a structured way. Assuming the chain is in a particular state i at time $t = 0$, the time (dwell time) spent in the initial state i must have a memoryless property according to one of the Markov properties, as discussed before. During a continuous-time process, the time between states has an exponential distribution that depends only on the i state.

The first step to build the Markov model is to estimate the intensity matrix ($\mathbf{Q}$), which can be initially calculated by Equation (2) using the historical record of the condition states assigned during inspections.

$$\mathbf{Q} = \begin{bmatrix} -\theta_1 & \theta_1 & 0 & 0 & 0 \\ 0 & -\theta_2 & \theta_2 & 0 & 0 \\ 0 & 0 & -\theta_3 & \theta_3 & 0 \\ 0 & 0 & 0 & -\theta_4 & \theta_4 \\ 0 & 0 & 0 & 0 & 0 \end{bmatrix} \; ; \; \begin{bmatrix} \theta_1 \\ \theta_2 \\ \theta_3 \\ \theta_4 \end{bmatrix} = \begin{bmatrix} q_{54} \\ q_{43} \\ q_{32} \\ q_{21} \end{bmatrix} \; ; \; q_{ij} = \frac{n_{ij}}{\sum \Delta t_i} \tag{2}$$

where q_{ij} represents the transition rate between adjacent states, n_{ij} is the number of elements that moved from state i to state j, and $\sum \Delta t_i$ is the sum of time intervals between observations, whose initial state is i.

The transition matrix ($\mathbf{P}$) is related to matrix $\mathbf{Q}$ through the following differential equation:

$$\frac{\partial \mathbf{P}}{\partial t} = \mathbf{PQ} \tag{3}$$

Equation (3) is known as the Chapman–Kolmogorov equation. Solving Equation (3), the transition matrix ($\mathbf{P}$) is given by the following expression:

$$\mathbf{P} = e^{\mathbf{Q}\Delta t} \tag{4}$$

In order to improve the quality of fit, the initial Markov model is improved through an optimization process, by minimizing Equation (5) [17]:

$$log\text{-}likelihood = \sum_{n=1}^{N} \sum_{m=1}^{M} \ln (\mathbf{P}_{ij}) \tag{5}$$

where M is the number of transitions observed in an element, N is the number of analyzed elements, and $\mathbf{P}_{ij}$ is the probability of occurrence of the observed transition, as predicted by the Markov model.

3.2. Artificial Neural Networks Models

ANN models have, as their primary source of inspiration, biological neural networks by the attempt to mimic the human brain's ability to recognize, associate, and generalize patterns. Ref. [19] defines the neural network as a massive parallel distributed processor consisting of simple processing units, which have the natural propensity to store experimental knowledge and make it available for use.

ANN is a nonlinear statistical technique capable of solving complex problems, able to learn and, therefore, to generalize. Generalization refers to the fact that the neural network produces adequate outputs for inputs that were not present during training. These information-processing capabilities enable neural networks to solve complex problems [19].

The ANN prediction model can be designed to predict the state condition of highway bridges. In this work, it was used a multiclass classification neural network (Figure 2) to develop the ANN prediction model. The Python language [20] and Scikit-learn package [21] were used to construct the ANN prediction model.

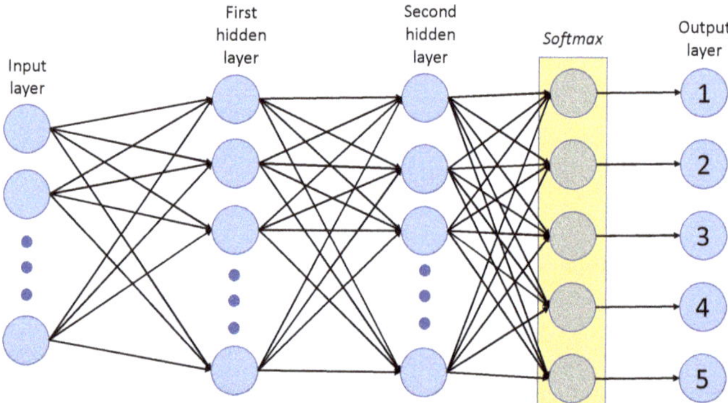

Figure 2. Multilayer perceptron with two hidden layers and softmax activation function.

All neurons in the hidden layers use the hyperbolic tangent activation function. The output layer uses the softmax function (Equation (6)). The softmax regression is a logistic regression that normalizes an input value into a vector that follows a probability distribution that totals up to 1. Additionally, as the asset cannot self-improve, an addition function was created to recalculate the probability vector output, imposing a 0 probability to any probability output that improves the condition state of the asset. Thus, the output is a vector $[P(5), P(4), P(3), P(2), P(1)]$, which presents the probability future condition states 5, 4, 3, 2, and 1, respectively.

$$\sigma(\overrightarrow{z})_i = \frac{e^{z_i}}{\sum_{j=1}^{K} e^{z_j}} \tag{6}$$

where, σ is the softmax function, $\overrightarrow{z}$ is the input vector, e^{z_i} is the standard exponential function for each element of the input vector, K is the number of classes in the multiclass classifier, and e^{z_j} is the standard exponential function for the output vector.

3.3. Statistical Tests

The statistical analysis of a model obtained in a given study is essential for validating and ensuring an acceptable extrapolation obtained for the population studied. Many statistical fit tests are applied to categorical data to assess how likely any observed difference happens by chance between the model and the observed data.

Many statistical tests can evaluate a model against the observed data. A commonly used statistical test is the chi-square fit test [22]. The chi-square fit test is applied to assess the fit between a set of observations (sample) and a theoretical distribution, comparing the distribution of sample data with the theoretical distribution to which the sample is supposed to belong. The test is an overall measure of the discrepancy between the frequencies observed in the sample and the expected frequencies (Equation (7)).

$$\chi^2 = \sum_{i=1}^{n} \frac{(O_i - E_i)^2}{E_i} \tag{7}$$

where χ^2 is the chi-square value, n is the total number of cells in the contingency table, E_i is the expected frequencies, and O_i is the observed frequencies.

The chi-square test limitation is to not consider the uncertainty inherent to the variable's possible outcomes. As the database presents a high number of bridges that remain in the same condition (from one inspection to another), using the chi-square test would get good results even with a DummyClassifier model (that consistently predicts the majority outcome [23]).

Cross-entropy can overcome this problem by measuring the dissimilarity between the sample data and the model. The cross-entropy evaluates the model on a test set to assess how accurate (based on the computation of likelihood) the model is in predicting the test data by calculating the "uncertainty" (or "information") of possible outcomes [24]. The entropy (H) is the expected value of "information" and is evaluated using the following Equation (8):

$$H(X) = E[I(X)] = -\frac{1}{N}\sum_{i=1}^{N} \ln p(x_i) \tag{8}$$

where X is the test set, E is the expected value operator, I is the information content of X, N is the size of the test set, and $p(x_i)$ is the probability of the predicted value x_i from the model.

Although entropy is crucial in assessing the quality of a model, the inverse of the perplexity (IPP) (Equation (9)) will be used to assess generalization performance. The inverse of the perplexity represents the probability of generating the expected outcome and has to be maximized. In other words, better models will tend to assign higher probabilities to the testing set. In this way, a DummyClassifier model would obtain zero as a result.

$$IPP(X) = \frac{1}{PP(X)} = \exp(-H(X)) \tag{9}$$

where IPP represents the inverse of the perplexity and PP symbolizes the perplexity.

With the statistical test defined, the next step is to distinguish between the model's performance on the training data and unseen test data. The common practice to avoid the model's overfitting is to divide the dataset into a training set (80%) and a test set (20%). Subsequently, a stratified k-fold cross-validation technique is performed using the training dataset (Figure 3), preserving the percentage of samples for each CI. The resampling method uses different portions of the data (A, B, C, and D) to train and validate a model on different iterations. The cross-validation gives an idea of how the model might perform in the worst-case and best-case scenarios when applied to new data [23]. To summarize the models' cross-validation accuracy, the mean (μ) and standard deviation (σ) are computed.

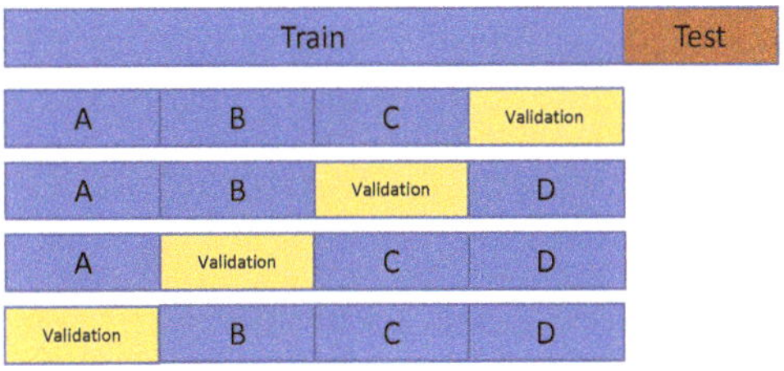

Figure 3. Train/test split and four-fold cross-validation.

4. Database

Due to data availability, and considering the large stock of bridges in different exposure conditions, Brazil was chosen as the object of study for this work. Brazil has an area of about 8,500,000 km^2 distributed in five regions with diverse climatic and social conditions, presenting an expressive group of bridges, with about 120,000 bridges [11].

A detailed inventory of 10,331 bridges was gathered up to the year 2021. The database includes information on the: geographic location, total length, deck width, material type, superstructure type, abutment type, standard traffic load model, year of construction, Average daily Traffic (ADT), Average daily truck traffic (ADTT), concession type, and structural condition.

The amount of information of the dataset varies from bridges presenting only their name, location, total length, and width to bridges with more detailed information, including results of inspections carried out, sketches, and photos.

The geographic coordinates of each bridge were exported to a Geographic Information System application [25]. Figure 4 shows the distribution of bridges, considering Brazilian geography, regions, and states. This figure clearly shows that most bridges (64%) are located in the southeast and northeast regions. According to the inventory collected, the states with highest number of bridges (above 500) are the states of Minas Gerais (MG), São Paulo (SP), Rio de Janeiro (RJ), Rio Grande do Sul (RS), Bahia (BA), Pernambuco (PE), Paraíba (PR), and Santa Catarina (SC). The states of MG (13.5%) and SP (12.3%) contain 26% of the Brazilian bridges.

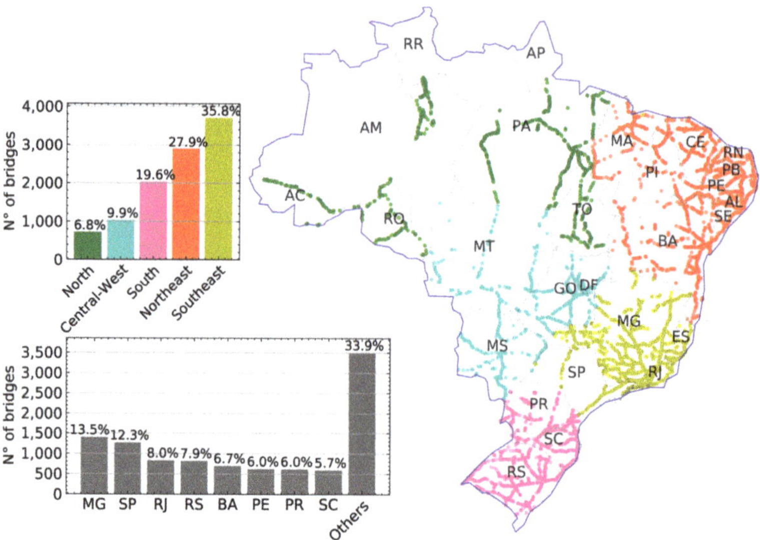

Figure 4. Distribution of bridges in Brazil.

Some key categories were analyzed and presented to understand the available data about the bridges. Regarding the material type, Figure 5a shows the number of bridges built per each category. It is possible to notice that the highest number of bridges are Reinforced concrete (RC), followed by Prestressed concrete (PC), amounting to about 71% of the total bridge inventory. As only some bridges are constructed by other materials (such as wood and masonry), they were clustered in one group (others). It is essential to observe that around 25% of the bridges do not have a material classification available in the database. Bridges' type represents the structural system of the bridge. As shown in Figure 5b, beam (54.4%) has the highest number among all bridges' types and slabs are the second (12.4%). As just a few bridges have a different structural system (such as arch bridge, cable-stayed bridge, suspension bridge, and truss bridge), they were clustered in one group (others). This work did not consider the number of spans, length of the largest span, static scheme, or skew angle, among others parameters, due to the scarcity of these data.

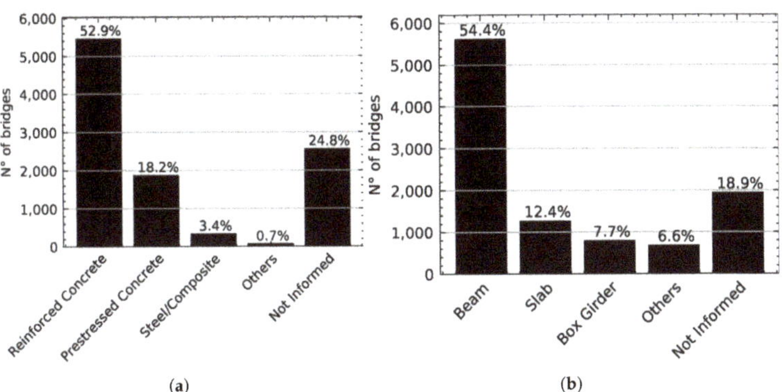

Figure 5. Distribution of bridges: (**a**) material type and (**b**) bridges' type.

The jurisdiction defines whether the bridge is managed by the federal, state, or municipal government. Figure 6a presents the number of bridges per different jurisdiction. The inventory shows that the federal administration is responsible for the most significant percentage of bridges (89.5%), followed by state administration (10.5%). Additionally, federal, state, and municipal administrations are subdivided into public and private concessions. Figure 6a shows that most bridges (66%) are under public concession. Even though the number of Federal public bridges in Figure 6a is way higher, the number of inspections of Federal private bridges (Figure 6b) overcomes the Federal public one. Considering inspections should happen once a year [13], it is expected that the total number of inspections would be proportional to the number of bridges and their ages, but this is not observed, showing that private concession seems to maintain the inspection schedule more strictly.

Figure 7 shows the distributions of the number of bridges per year of construction. It is observed that 63% of Brazilian bridges are over 50 years of age. It also manifests the periods of most significant and minor investment in the bridge construction sector, highlighting the positive impact of the Maurício Joppert Law (1945), the period of President Juscelino Kubitscheck (1956 to 1960), the revolutionary period started in 1964, in particular the construction of the Rio-Niterói Bridge (1974), the negative impact of the 1988 Constitution that radically altered the financing sector, and the resumption of investment during the period of President Fernando Henrique Cardoso (1994 to 2002) [9]. Additionally, the ages of the bridges allow estimating the live load. For example, concerning the standard traffic load model, most bridges designed between 1960 and 1975 considered a TB-36 on roads: a vehicle with 360 kN force [26]. The bridges designed after 1985, with the new version of the Brazilian standard [27], started to consider vehicles with 450 kN force (TB-45) as standard traffic load model. The high number of bridges for which the year of construction is unknown (or was not informed) is just one example of the difficulties found for a more detailed analysis of the existing situation and evidence of missing record information.

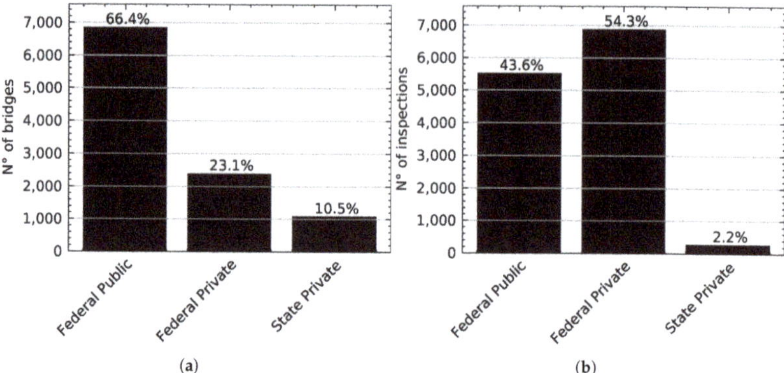

Figure 6. Distribution of bridges (**a**) and inspections (**b**) per jurisdiction and concession type.

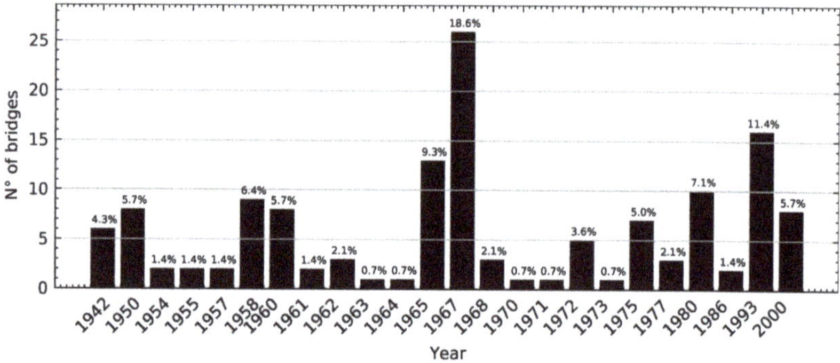

Figure 7. Number of bridge by year of construction.

The ADT is the average volume of traffic recorded in one day (24 h). This data is used to assess traffic distribution, measure the demand for a road, and schedule basic improvements. Figure 8 presents histogram distribution of the ADT and ADTT. The histogram clearly shows that ADT less than 10,000 has the highest frequency. For the ADTT case, traffic less than 2000 has the highest frequency. It can also be seen that the ADTT corresponds, on average, to 34% of ADT.

The bridges sum up approximately 626 km of length distributed according to Figure 8c, with an average of 69 m. The data shows that 40.3% of the bridges have an extension equal to or less than 30 m, and 29.0% show an extension higher than 60 m.

Visual inspections constitute the main form of evaluation of bridges in Brazil, and the records of these inspections are used to assist in the construction and history of the bridge inventory. The first inspections presented in the inventory date back to 2008, covering approximately 13 years and resulting in 24,127 inspections. Figure 9 shows the periodicity of inspections carried out in the period mentioned above. The time between inspections that equals 0 means that only one inspection is available for the bridge. As can be concluded, visual inspections are carried out mostly annually, according to what was discussed in Figure 2.

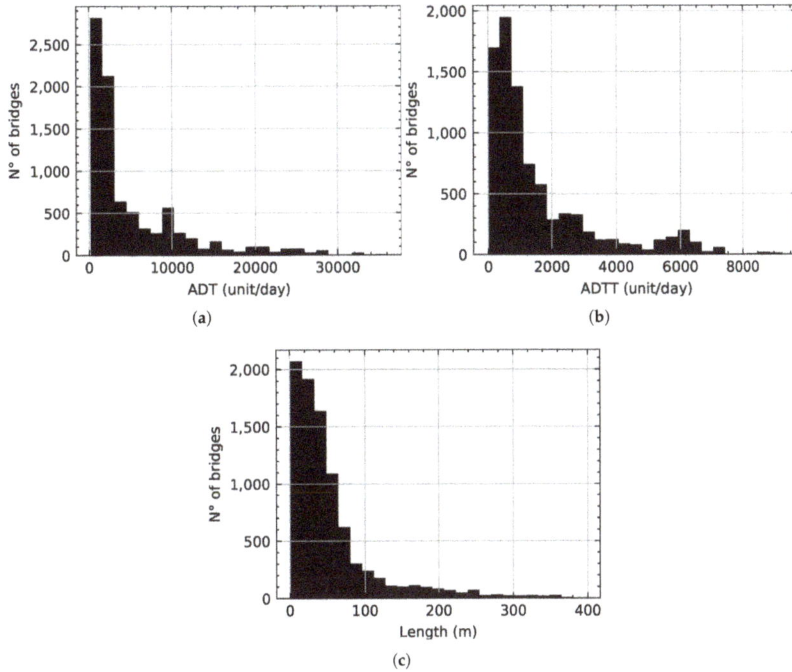

Figure 8. Distribution of bridges according to (**a**) ADT, (**b**) ADTT, and (**c**) total length.

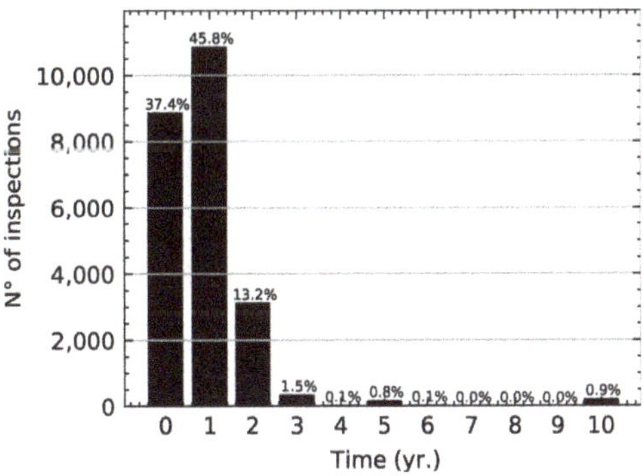

Figure 9. Periodicity of inspections.

5. Methodology and Results

The previously described bridges' condition data are used to develop the Markov model and the ANN model. Before implementing the predictive models, a filtering process was performed on the database to remove inconsistencies. For example, ungraded records were removed, along with cases where an improvement in grade was observed. This last effect can be attributed to maintenance actions not included in the inventory or imprecision in evaluating the bridge's condition due to the subjectivity of the inspectors in the visual inspection technique [28–30]. Unfortunately, the database does not differentiate one case from the other, and all positive transitions had to be assumed as repair activities [30].

The original database includes a total of 10,331 bridge assets and 24,127 inspections. After the filtering process, only 7,754 bridges and 12,681 inspections were considered for the deterioration modeling, keeping a reasonable amount of data [28,31]. A summary of the inspection transitions are presented in Figure 10.

	5	4	3	2	1
5	681	382	28	5	2
4	0	6196	824	62	14
3	0	0	3182	194	33
2	0	0	0	897	59
1	0	0	0	0	122

Figure 10. Matrix transition indexes.

5.1. Markov Model

As bridges in a network are likely to share comparable environmental conditions, it would be plausible to presume that the deterioration processes should be similar in equivalent scenarios if considering load and environment. Hence, in the first part of the analysis, the bridges were assumed to have the same deterioration processes regardless of structure type, location, and material of the superstructure.

Equation (10) shows the developed transition probability matrices for a general bridge, representing the deterioration process under normal operational conditions in Brazil. Table 3 summarizes the Markov model's accuracy. The model achieved the same accuracy during training and validation but had a higher dispersion during the latter. The model reached a slightly lower accuracy during testing but showed that it could generalize to the unseen data.

$$\begin{bmatrix} \theta_1 \\ \theta_2 \\ \theta_3 \\ \theta_4 \end{bmatrix} = \begin{bmatrix} 0.42326519 \\ 0.09922577 \\ 0.0597748 \\ 0.0695101 \end{bmatrix} \qquad \mathbf{P}(1) = \begin{bmatrix} 0.655 & 0.3274 & 0.017 & 0.0004 & 0.00 \\ 0 & 0.9055 & 0.092 & 0.003 & 0.00 \\ 0 & 0 & 0.942 & 0.056 & 0.002 \\ 0 & 0 & 0 & 0.933 & 0.067 \\ 0 & 0 & 0 & 0 & 1 \end{bmatrix} \qquad (10)$$

Table 3. Markov model performance during training, validation, and testing.

Dataset	IPP $\mu(\sigma)$
Training	0.670 (0.003)
Validation	0.670 (0.012)
Testing	0.664

The following results are presented as a form of illustration considering a 100-year time horizon for a general bridge asset. During this period, no maintenance activities are assumed, i.e., the bridge is allowed to decay constantly.

Figure 11a shows the probability of a bridge to be in a determined CI over time. It is possible to observe that the probability of a bridge remaining in the CI equal to 5 drops significantly in just a few years. For the case of the CI equal to 3 and 2, it is possible to observe a more flat curve, meaning that these CIs remain for a longer period. Brazilian standards do not specify a lifespan for the structures, but in principle, it seems to be assumed as 50 years. Considering 50 years as the lifespan, 70% of bridges reach the CI equal to 1. In Figure 11b, the average of the bridges reaches the CI equal to 3 in 14 years and the CI equal to 2 in 32 years. Additionally, it is possible to observe a high dispersion in the time that the bridge reaches the CI equal to 2, ranging from 10 to 75 years.

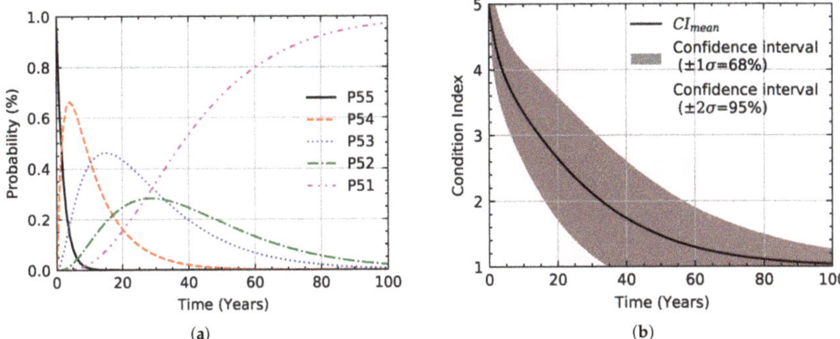

Figure 11. Results from Markov model. (**a**) Transition probability over time; (**b**) expected condition over time.

Clusters

Previously, the bridges were assumed to have the same deterioration processes. However, it is plausible that the deterioration processes should be different in bridges subjected to different conditions, such as material type, geographic location, and geometric properties. Thus, taking a step forward to improve the results, the bridges were clustered into specific categories.

The first attempt of clustering the database was by material type. Thus, considering the types of materials mentioned before (Figure 5a), the Markov model was run considering only the bridge records belonging to each material type. Figure 12a presents the deterioration of bridges with different materials. As observed, reinforced and prestressed concrete bridges have similar deterioration rates, whereas steel and composite bridges have a higher deterioration.

Concerning geographic location, the bridges were categorized by states and regions (Figure 4) to demonstrate their different performance and lifespan. Figure 12b shows the deterioration curves for each state, showing no significant difference between states from condition 5 to condition 3. If condition 2 (poor condition in Table 1) is adopted as the minimum acceptable condition, the predicted average service life of a bridge in BA, RJ, and SP is 25, 45, and 60 years, respectively, with SP presenting the best performance. The anomalous behavior of the SC curve is justified by a high density of inspections with CI equal to 5 in 2015 and 2016, having no further inspections after it. The significant variation of the bridge service life illustrates the considerable impact of the state management policies on the performance of a bridge.

Figure 12c presents the results obtained regarding the regions, where it is possible to observe no significant difference between them, except the central–west region. The only remark, in this case, is the small amount of data that might hinder the actual behavior of this region compared to the others.

Additionally, the bridges were grouped into three categories according to their length. As discussed in Section 4, the bridges were rated as short (for bridges shorter or equal 30 m), medium (for bridges longer than 30 m and shorter than 60 m), and long (for bridges longer than 60 m). As illustrated in Figure 12d, longer bridges show a higher deterioration compared to medium and short bridges. This observation is reasonable as it is expected that there is a higher probability of having deterioration in a long bridge than in a short one.

The clustered Markov models reached higher accuracy during testing than the not clustered one, as summarized in Table 4. It can be observed that the most remarkable improvement was achieved when splitting the data by material type. These results show that adding new features to the model help to improve its performance.

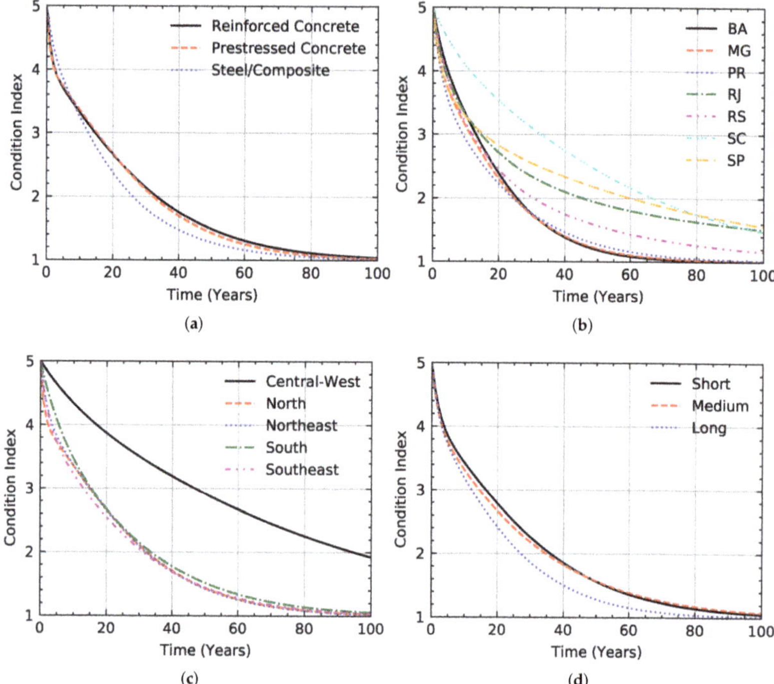

Figure 12. Expected condition over time. (**a**) Material type; (**b**) state; (**c**) region; (**d**) bridge length.

Table 4. Clustered Markov models performance during testing.

Cluster	*IPP*
All	**0.664**
Material	0.683
State	0.667
Region	0.672
Length	0.679

This section has evidenced the influence of selecting a suitable cluster to predict a typical bridge performance. As these predictions are the basis for BMS, inaccuracies might have significant impacts, especially in budget plans. Additionally, carrying out analysis with a fragmented database, as in the case of the Markov model, does not seem to be a good alternative as it can lead to unsafe predictions.

5.2. Artificial Neural Network

Similar to the Markov model analysis, the first assumption was to consider similar deterioration processes for all bridges. For the first ANN model, two input nodes were considered in the input layer, one for the initial CI and the other for the time between inspections (Figure 9). The optimal configuration for the ANN classifier was determined by trial and error. The inverse of the perplexity (*IPP*) was used as an entropy indicator, where a higher rate indicates a lower entropy between the labels and predictions. Accordingly, the final configuration of the ANN is a one-layer network with five hidden neurons in the layer.

Comparing the ANN model against the Markov model, it is possible to observe, in Table 5, that the ANN model achieved higher accuracy during training, validation, and testing. Even though the test accuracy in the ANN model is slightly lower than the training and validation, the results show that the model could generalize.

Table 5. ANN and Markov models performance during training, validation, and testing.

Model	Training	Validation	Testing
Markov	0.670 (0.003)	0.670 (0.012)	0.664
ANN	0.673 (0.003)	0.673 (0.010)	0.670

With the model calibrated, the performance of the bridge can be evaluated over the years. Three approaches were considered to predict the model throughout the years. For the first approach (Figure 13a), it is reasonable to assume the extrapolation of up to 100 years. As can be seen, the model could not extrapolate or make predictions outside the training data range. Once it leaves the range for which the model has data, it simply keeps predicting the last known point until it reaches a specific one (40 years), where it drops abruptly. The model cannot generate "new" responses outside of what was seen in the training data.

Another approach to predicting performance over the years is to obtain the transition probability matrix for one year and evaluate the performance over the years, similar to the Markov model. Figure 13b presents the results for this approach. It is possible to observe that the ANN model yields higher deterioration results. However, it is tricky to consider only one year between inspections when working with the ANN model as it was trained with a time distribution of up to 10 years (Figure 9).

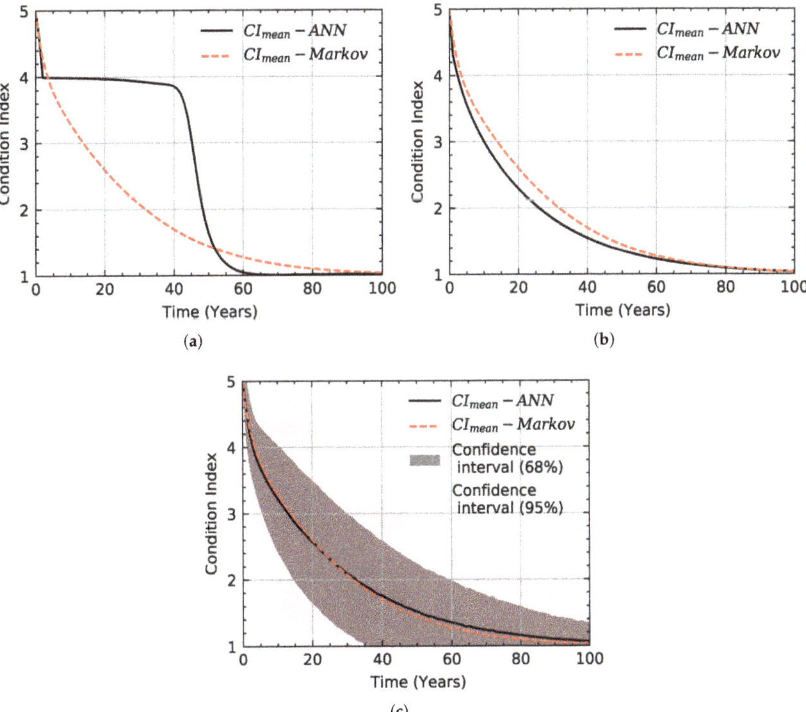

Figure 13. Evolution of the degradation over time by ANN model. (**a**) Extrapolation approach; (**b**) 1 year approach; (**c**) Monte Carlo approach.

The final approach considered to obtain the CI through the years was a Monte Carlo (MC) simulation based on the inspection periodicity distribution in Figure 9. During the simulation, one million artificial bridges were generated with an initial CI equal to 5. For each bridge, a random inspection periodicity was picked from the distribution (Figure 9) and a new CI was calculated based on the probability from the calibrated model. After all the bridges achieved 100 years, the mean and standard deviation were calculated by year, generating the expected values and confidence interval presented in Figure 13c. As can be seen, the results from the ANN are now very similar to the Markov results.

5.2.1. Filling the Database by ANN

As it was pointed out, the inspection periodicity as an input parameter can increase the complexity of the ANN model. Therefore, a reasonable strategy is to eliminate it as input by filling the empty inspections in the dataset and making the interval between inspections constant (1 year). Different approaches could be used to reach such desired results. One could assume the bridge maintains the same CI until the next inspection. For example, if a bridge has a CI equal to 4 in 2008 and in the next inspection, in 2011, its CI is equal to 2, the CI assumed for 2009 and 2010 would be 4 for both years. Another approach would be to use a linear interpolation between the inspection records to fill the gap. In this case, the CI for 2009 and 2010 would be 3 and 3.

Even though these approaches could solve the problem, they do not consider the stochastic behavior of the deterioration process. Thus, predictive models such as Markov and ANN would be a better option to fill in the gaps reliably. For the context of this work, and considering the results obtained in Table 5, the approach selected is the one proposed by [32], where the authors used an ANN model to generate artificial historical bridge condition indexes. The example already discussed was used to exemplify the application of the selected methodology, as illustrated in Figure 14.

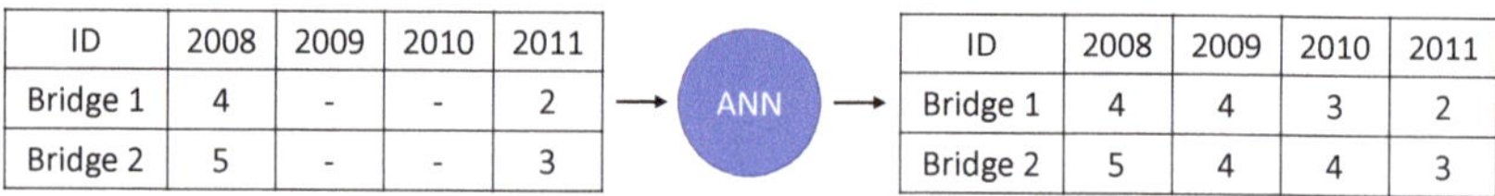

Figure 14. Filling by ANN.

The final dataset was filled by artificial historical bridge condition indexes generated from a MC simulation using the proposed methodology [32] and the inspection periodicity distribution in Figure 9. A total of 4720 data points were generated, inducing the database to go from 12,681 to 17,401 inspections.

5.2.2. Additional Features

Considering the deterioration processes should differ in bridges subjected to different conditions, the model was updated by adding new features one at a time, attempting to improve the results. The new parameters found to be significant to bridge deterioration were the Gross domestic product (GDP), ADTT, total length, and geographic location. For the case of categorical data, one-hot encoding [21,23] was used to feed the model. Figure 15 presents the results for some of the new parameters.

Previously, in the Markov model, the total length was modeled as a categorical parameter. Now, for the case of the ANN model, it has been modeled as a continuum parameter. The results presented in Figure 15a confirm the same outcome presented in Figure 12d, where longer bridges show a higher deterioration than shorter bridges. It is important to point out here the advantage of the ANN model, over Markov models, in being able to use continuum inputs instead of only categorical ones.

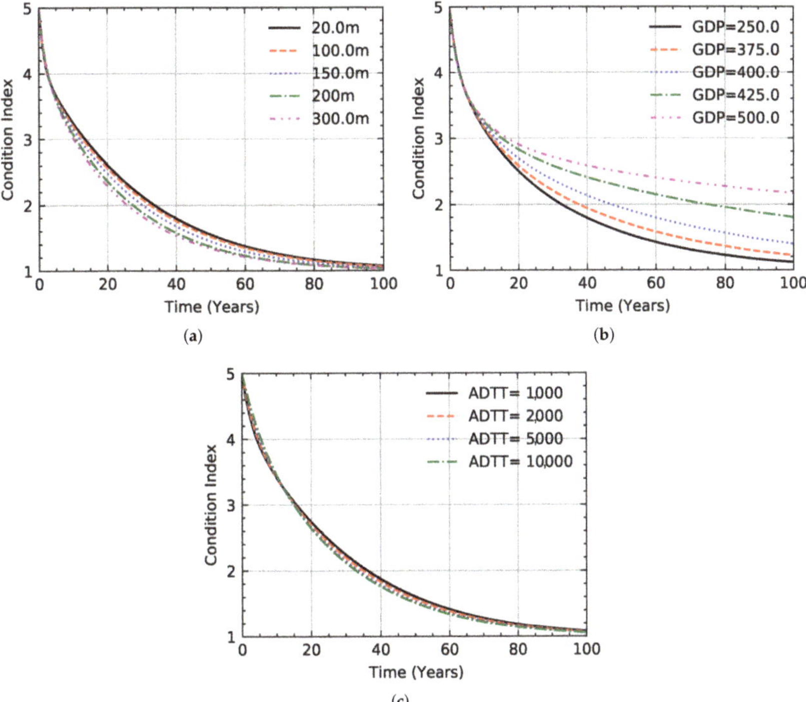

Figure 15. Expected condition over time. (**a**) Bridge length; (**b**) GDP in USD billion for 2018; (**c**) ADTT.

The results presented earlier in Figure 12b clarify and help to understand how each state manages its bridges. However, the state is a categorical feature with no natural meaning built into it. In order to view the situation from another perspective, the categorical parameter (states) was converted into an economic continuum parameter (GDP). Figure 15b illustrates the influence of the GDP. It is possible to observe that states with a higher GDP have a better performance. The significant variation of the bridge service life demonstrates the considerable impact of the GDP on the performance of the bridge. To have an idea of the impact, let us adopt condition 2 (poor condition in Table 1) as the minimum acceptable condition. By observing Figure 15b, it is possible to notice that it could vary from 30 years up to 100 years. Additionally, no significant difference from condition 5 to condition 3 was observed.

Some studies [33,34] have observed a significant influence of ADTT on bridge performance, by an negative correlation. Thus, it was expected to observe the same results in the analysis carried out. However, as demonstrated in Figure 15c, the results show a low, or almost zero, influence of ADTT on bridge performance. A possible explanation of why the ADTT did not present the expected results is the fact that the ADTT and the GDP have a positive correlation. In other words, the influence of the GDP might be hiding the influence of ADTT. Another possible reason is that bridges with a high ADTT usually receive more attention from the managers. Further analysis considering the weight and traffic distribution [35] could help to enhance the results.

The knowledge of the degree of atmospheric aggressiveness is vital in building maintenance management to ensure the project's useful life [36]. Humidity and high temperatures notably favor the degradation processes of materials exposed to the atmosphere. Wetting time, type and concentration of gaseous pollutants and particulate matter in the atmosphere determine the magnitude of the attack. The availability of values for these variables greatly assists in assessing the potential risk of corrosion. Figure 16 illustrates a modified Brooks atmospheric corrosivity index for the Brazilian region.

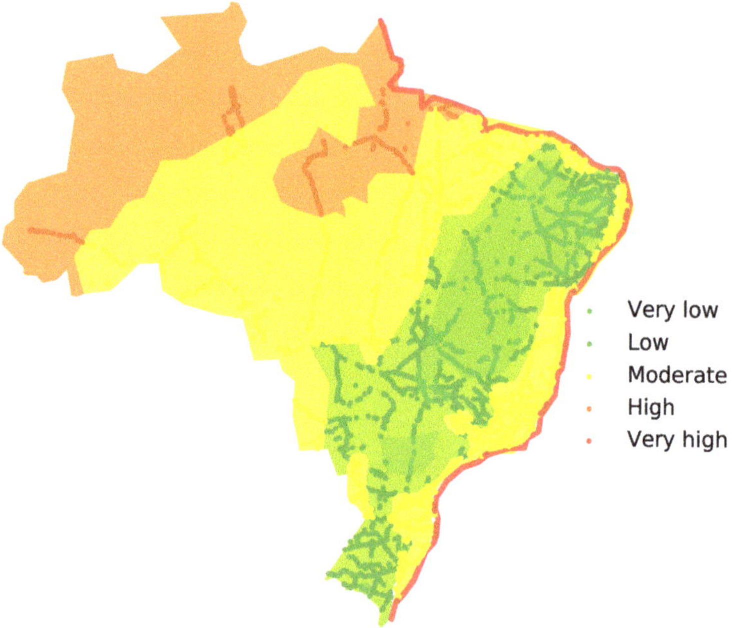

Figure 16. Modified Brooks atmospheric corrosivity index (adapted from [36]).

Although the theoretical conceptualisation in the Brooks index is straightforward, as it only considers humidity and temperature as intervening factors, it can be of value in qualifying the aggressiveness of large rural areas in Brazil, where the data are scarce or nonexistent. On the other hand, when evaluating large cities, industrial areas, and coastal regions, it is essential to be aware that the polluting agents (not considered in the Brooks index [36]) cause a substantial increase in the corrosion rate.

Considering that the location of bridges influences the speed of bridge degradation, particularly for bridges located on the coastal strip, different distances from the coastline were analyzed. Figure 17a presents the results for coastline distance. As can be observed, the coastline distance significantly influences the deterioration process. Figure 17b illustrates the influence of each aggressive zone, identifying an additional band (very high) into the map proposed by Brooks (Figure 16), defined by a distance of less than 5 km from the coast. From Figure 16, it is possible to conclude that bridges located on the coastline and in the north of Brazil have an unquestionably high deterioration process.

As we are now using a new database, a new analysis not considering any cluster was performed to be used as a reference value. A summary of the results is presented in Table 6. These results show that adding new features to the model help to improve its performance.

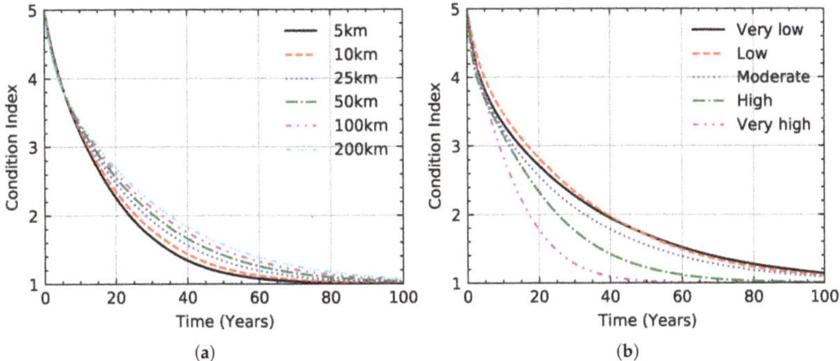

Figure 17. Expected condition over time: (**a**) Coastline distance; (**b**) aggressive zones.

Table 6. Clustered ANN models performance during testing.

Model	Training	Validation	Testing
All	0.729 (0.005)	0.729 (0.002)	0.710
Length	0.744 (0.009)	0.743 (0.002)	0.718
GDP	0.733 (0.005)	0.733 (0.003)	0.712
ANTT	0.736 (0.009)	0.736 (0.002)	0.718
Aggressive zones	0.735 (0.005)	0.732 (0.002)	0.713

5.2.3. Proof of Concept

A good-quality control plan specifies the extent and the interval of inspections and the data necessary to estimate performance indicators and forecast future development [4]. In this context, planning is essential to establish a schedule, scope, and optimal times between inspections. As discussed in Section 2, visual inspection assessment practice differs from standard to standard [12–14] when defining the frequency of inspections (periodicity).

The uniform interval approach has resulted in a very costly and inefficient process [37]. Provisions for adjusting the frequency of routine inspection for certain types or groups of bridges to better conform with their inspection needs have been defined by [38], taking into account the actual condition index, length, load redundancy, susceptibility to damage, structure type, maintenance history, structure age, ADT, and ADTT. Ref. [1] developed a framework for risk-based bridge inspection that identifies bridges for which inspection intervals shorter or longer than the one defined by the standards are more appropriate.

According to [4], the frequency of bridge inspections should depend on bridge condition and bridge importance to the network. Therefore, bridges with poor condition and the most critical bridges should be inspected more frequently than most bridges in the network. On the other hand, new(er) bridges with little or no damage could be inspected less frequently. Additionally, bridges with different material characteristics and locations may require different attention. In order to answer this, bridges with different material characteristics and locations are studied, and their performance is compared.

Six representative groups were selected to represent the case studies, with one bridge representing each corresponding group. Three corrosivity zones (Figure 16) and the two main materials (Figure 5) were selected, as shown in Table 7.

The predicted performance and service life of bridges for each representative group (Table 7) are presented in Figure 18. As it can be seen, the bridges have a significant distinction in performance. RC bridges located in a very high corrosive zone have an elevated deterioration process. By adopting condition 2 (poor condition in Table 1) as the minimum acceptable condition, the predicted average service life for bridges located in the very high corrosive zone is around 20 years, whereas it is 40 years for a very low corrosive zone. The significant variation of the bridge service life illustrates the considerable impact of the corrosive zone on the performance of the bridge, making it almost mandatory to consider the aggressiveness of the region where the bridge is located when defining the periodicity of inspections.

Table 7. Selected representative bridge groups.

Name	Aggressive Zone	Material
Bridge-1	Very low	RC
Bridge-2	Very low	PC
Bridge-3	Moderate	RC
Bridge-4	Moderate	PC
Bridge-5	Very high	RC
Bridge-6	Very high	PC

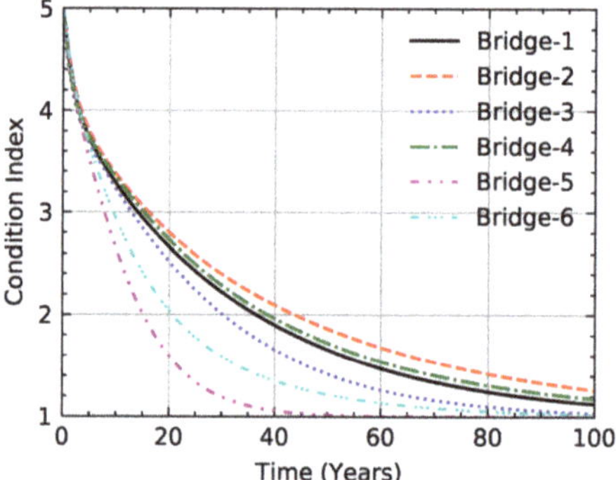

Figure 18. Expected condition over time by representative bridge groups.

6. Conclusions

The primary contribution of this study to BMS is the application of stochastic models to probabilistically forecast bridge deterioration and the execution of a systematic method to show new features importance in the degradation process, resulting in exciting insights into the definition of inspection periodicity.

The Brazilian standards have different values related to the interval between inspections, but they all consider the periodicity constant, not giving consideration to the bridges' singularities. As discussed throughout this article, bridges in a network are likely to share similar environmental conditions but, depending on their age, location, structural typology, and other aspects, they may be exposed to different structural deterioration process over time. Hence, forecasting simulations were carried out to identify the bridge behavior for different scenarios.

Two predictive models (Markov and ANN) were created to predict future bridge conditions based on historical data. The most representative up-to-date database of the construction site was served as input for the models, containing information about 10,331 bridges in Brazil from 2008 to 2021.

Considering the deterioration curve obtained for the whole dataset in Figures 11b and 13c, the bridges will have, at the end of their 30 years, a condition rating of around 2, if only routine maintenance is performed. Additionally, the forecasting results of two predictive models (Markov and ANN) indicate that the ANN model can predict future conditions more accurately than the Markov one.

In this work, some clusters were identified to improve existing BMS, especially atmospheric corrosivity, which has a significant influence on the deterioration process. The results obtained indicate the proposition of a variation in the periodicity of inspections as a function of the bridges' degradation curves. While this investigation addressed limited features, other continuous and categorical variables can be added to the methodology (such as the wearing surface and skew angle), which could improve the prediction accuracy of the methodology.

It is essential to emphasize that, regardless of the outstanding potential of Markov and ANN models, the bridge engineer's opinion must not be ignored. On the contrary, when an expert validates the obtained results, they gain plausibility and can be used confidently.

There is a need to expand and strengthen the works' inventory to calibrate the results obtained. To achieve this goal, it is necessary to implement joint efforts from all managers, industry professionals, and researchers linked to bridge engineering to promote sharing information, enabling this work to be expanded nationally and internationally.

One limitation of the work is that the results obtained to improve the frequency of inspections are limited to internal factors (deterioration) and do not consider extreme events, such as floods, earthquakes, or any vandalism act.

Author Contributions: Conceptualization, A.F.S.; methodology, A.F.S., M.S.B. and H.S.S.; software, M.S.B.; validation, H.S.S., T.N.B. and J.C.M.; formal analysis, M.S.B.; investigation, A.F.S., M.S.B. and H.S.S.; data curation, A.F.S. and M.S.B.; writing—original draft preparation, A.F.S. and M.S.B.; writing—review and editing, A.F.S., M.S.B., H.S.S., T.N.B. and J.C.M.; visualization, A.F.S. and M.S.B.; supervision, H.S.S., T.N.B. and J.C.M.; project administration, H.S.S., T.N.B. and J.C.M.; funding acquisition, H.S.S. and J.C.M. All authors have read and agreed to the published version of the manuscript.

Funding: This project has received funding from the European Union's Horizon 2020 research and innovation program under grant agreement No 769255. This document reflects only the views of the author(s). Neither the Innovation and Networks Executive Agency (INEA) nor the European Commission is in any way responsible for any use that may be made of the information it contains. This work was partly financed by FCT / MCTES through national funds (PIDDAC) under the R&D Unit Institute for Sustainability and Innovation in Structural Engineering (ISISE), under reference UIDB / 04029 / 2020.

Data Availability Statement: The data presented in this study are available on request from the corresponding author.

Acknowledgments: The authors thank the Brazilian National Department of Transport Infrastructure (DNIT) and the Brazilian National Land Transportation Agency (ANTT) for providing information that enabled the analysis of the studied structures.

Conflicts of Interest: The authors declare no conflict of interest.

References

1. Ilbeigi, M.; Pawar, B. A Probabilistic Model for Optimal Bridge Inspection Interval. *Infrastructures* **2020**, *5*, 47. [CrossRef]
2. Baron, E.A.; Galvão, N.; Docevska, M.; Matos, J.C.; Markovski, G. Application of quality control plan to existing bridges. *Struct. Infrastruct. Eng.* **2021**, 1–17. [CrossRef]
3. Santarsiero, G.; Masi, A.; Picciano, V.; Digrisolo, A. The Italian Guidelines on Risk Classification and Management of Bridges: Applications and Remarks on Large Scale Risk Assessments. *Infrastructures* **2021**, *6*, 111. [CrossRef]
4. Hajdin, R.; Kušar, M.; Mašović, S.; Linneberg, P.; Amado, J.; Tanasić, N. *WG3—Technical Report Establishment of a Quality Control Plan*; TU1406 Cost Action: Guimarães, Portugal, 2018.
5. Hadjidemetriou, G.M.; Herrera, M.; Parlikad, A.K. Condition and criticality-based predictive maintenance prioritisation for networks of bridges. *Struct. Infrastruct. Eng.* **2021**, 1–16. [CrossRef]
6. Ghonima, O. Statistical Modeling of United States Highway Concrete Bridge Decks. Ph.D. Thesis, University of Delaware, Newark, DE, USA, 2017.
7. Oliveira, C.B.d.L. Determinação e análise de taxas de Deterioração de pontes Rodoviárias do Brasil. Ph.D. Thesis, Federal University of Minas Gerais, Pampulha, Brazil, 2018.
8. Gomes, C.; Monteiro, E.; Vitório, A. A study on the structural degradation of road bridges and viaducts. In Proceedings of the XIII International Conference on Structural Repair and Rehabilitation, Crato, Brazil, 7–9 September 2017; pp. 1–15.
9. Mendes, P.d.T.C. Contribuição para um modelo de gestão de pontes de concreto aplicado à rede de rodovias brasileiras. Ph.D. Thesis, University of São Paulo, São Paulo, Brazil, 2009.
10. Oliveira, C.; Greco, M.; Bittencourt, T. Analysis of the brazilian federal bridge inventory. *IBRACON Estrut. Mater.* **2019**, *12*, 1–13. [CrossRef]

11. Santos, A. Alguém sabe quantas pontes existem nas rodovias do Brasil? 2017. Available online: http://www.guiadotrc.com.br/noticias/noticiaID.asp?id=36867 (accessed on 20 October 2021).
12. *DNIT 010*; Inspeções em pontes e viadutos de concreto armado e protendido—Procedimento. Departamento Nacional de Infraestrutura de Ttransportes: Rio de Janeiro, Brazil, 2004.
13. *ABNT NBR 9452*; Inspection of Concrete Bridges and Footbridges—Procedures. Associação Brasileira de Normas Técnicas: Rio de Janeiro, Brazil, 2019.
14. *ARTESP ETC 21-002*; Controle Das Condições Estruturais, Funcionais e de Durabilidade das Obras de Arte Especiais. Agência de Transportes do Estado de São Paulo: São Paulo, Brazil, 2007.
15. Srikanth, I.; Arockiasamy, M. Deterioration models for prediction of remaining useful life of timber and concrete bridges: A review. *J. Traffic Transp. Eng. (Engl. Ed.)* **2020**, *7*, 152–173. [CrossRef]
16. Adey, B.T.; Hajdin, R. Methodology for determination of financial needs of gradually deteriorating bridges with only structure level data. *Struct. Infrastruct. Eng.* **2011**, *7*, 645–660. [CrossRef]
17. Denysiuk, R.; Fernandes, J.; Matos, J.C.; Neves, L.C.; Berardinelli, U. A computational framework for infrastructure asset maintenance scheduling. *Struct. Eng. Int.* **2016**, *26*, 94–102. [CrossRef]
18. Kallen, M.J.; Van Noortwijk, J.M. Statistical inference for Markov deterioration models of bridge conditions in the Netherlands. In Proceedings of the 3rd International Conference on Bridge Maintenance, Safety, Management, Life-Cycle Performance and Cost, Porto, Portugal, 16–19 July 2006; pp. 1–8.
19. Haykin, S. *Neural Networks and Learning Machines*, 3rd ed.; Prentice Hall: Hoboken, NJ, USA, 2008.
20. Python Software Foundation. Python Language Reference, Version 3.9. 2015. Available online: http://www.python.org (accessed on 20 October 2021).
21. Pedregosa, F.; Varoquaux, G.; Gramfort, A.; Michel, V.; Thirion, B.; Grisel, O.; Blondel, M.; Prettenhofer, P.; Weiss, R.; Dubourg, V.; et al. Scikit-learn: Machine Learning in Python. *J. Mach. Learn. Res.* **2011**, *12*, 2825–2830.
22. Jackson, S.L. *Research Methods and Statistics: A Critical Thinking Approach*, 5th ed.; Cengage Learning: Belmont, CA, USA, 2009.
23. Müller, A.C.; Guido, S. *Introduction to Machine Learning with Python and Scikit-Learn*; O'Reilly Media: Sebastopol, CA, USA, 2015.
24. MacKay, D.J; Mac Kay, D.J. *Information Theory, Inference, and Learning Algorithms*, 1st ed.; Cambridge University Press: New York, NY, USA, 2003.
25. Jordahl, K.; den Bossche, J.V.; Fleischmann, M.; Wasserman, J.; McBride, J.; Gerard, J.; Tratner, J.; Perry, M.; Badaracco, A.G.; Farmer, C.; et al. Geopandas/Geopandas: V0.10.2. 2021. Available online: https://geopandas.org/ (accessed on 20 October 2021). [CrossRef]
26. *ABNT NBR 7188*; Cargas móveis em pontes Rodoviárias. Associação Brasileira de Normas Técnicas: Rio de Janeiro, Brazil, 1960.
27. *ABNT NBR 7188*; Road and Pedestrian Live Load on Bridges, Viaducts, Footbridges and Other Structures. Associação Brasileira de Normas Técnicas: Rio de Janeiro, Brazil, 2013.
28. Kotze, R.; Ngo, H.; Seskis, J. *Improved Bridge Deterioration Models, Predictive Tools and Costs*; Research Report; Austroads: Sydney, Australia, 2015.
29. Bu, G.; Lee, J.; Guan, H.; Blumenstein, M.; Loo, Y. Typical deterministic and stochastic bridge deterioration modeling incorporating backward prediction model. *J. Civ. Struct. Health Monit.* **2013**, *3*, 141–152. [CrossRef]
30. Thompson, P.D.; Ford, K.M.; Arman, M.H.R.; Labe, S.; Sinha, K.C.; Shirole, A.M. *Estimating Life Expectancies of Higway Assets: Volume 1: Guidebook*; NCHPER Report 13; Transportation Research Board: Washington, DC, USA, 2012; p. 78.
31. Tran, H. D Sensitivity of Markov Model to Different Sampling Sizes of Condition Data. *J. Perform. Constr. Facil.* **2016**, *30*, 06015005. [CrossRef]
32. Lee, J.; Sanmugarasa, K.; Blumenstein, M.; Loo, Y.C. Improving the reliability of a Bridge Management System (BMS) using an ANN-based Backward Prediction Model (BPM). *Autom. Constr.* **2008**, *17*, 758–772. [CrossRef]
33. Mauch, M.; Madanat, S. Semiparametric Hazard Rate Models of Reinforced Concrete Bridge Deck Deterioration. *J. Infrastruct. Syst.* **2001**, *7*, 49–57. [CrossRef]
34. Huang, Y.H. Artificial Neural Network Model of Bridge Deterioration. *J. Perform. Constr. Facil.* **2010**, *24*, 597–602. [CrossRef]
35. Portela, E.L. Analysis and Development of a Live Load Model for Brazilian Concrete Bridges Based on WIM Data. Ph.D. Thesis, University of São Paulo, São Paulo, Brazil, 2018.
36. Pannoni, F.D. *Manual de Construção em Aço—Projeto e Durabilidade*; CBCA: Rio de Janeiro, Brazil, 2017.
37. Nasrollahi, M.; Washer, G. Estimating Inspection Intervals for Bridges Based on Statistical Analysis of National Bridge Inventory Data. *J. Bridge Eng.* **2015**, *20*, 04014104. [CrossRef]
38. Federal Highway Administration (FHWA). *Revisions to the National Bridge Inspection Standards (NBIS)*; Technical Advisory 5140.21; FHWA: Washington, DC, USA, 1988.

MDPI

Article

How Long Can a Wood Flooring System Last?

Pedro Coelho [1], Ana Silva [2,*] and Jorge de Brito [2]

[1] Instituto Superior Técnico (IST), University of Lisbon, Av. Rovisco Pais, 1049-001 Lisbon, Portugal; pedro.abm.coelho@hotmail.com

[2] Department of Civil Engineering, Architecture and Georesources, Instituto Superior Técnico (IST), University of Lisbon, Av. Rovisco Pais, 1049-001 Lisbon, Portugal; jb@civil.ist.utl.pt

* Correspondence: ana.ferreira.silva@tecnico.ulisboa.pt

Abstract: Wood is a natural, sustainable, and renewable material, which has been used as flooring for centuries, but not enough is known about its durability and performance over time when subjected to different degradation agents. This study proposes a methodology for the service life prediction of wood flooring systems, considering the impact of different factors that influence the floors' durability. For that purpose, a fieldwork survey is performed to evaluate the degradation phenomena of 96 indoor wood floorings in-use conditions, located in Portugal. The data collected are converted into degradation patterns that graphically illustrate the loss of performance of wood floorings over time. An estimated service life of 44 years is obtained. This study thus allows quantifying the impact of various characteristics on the indoor wood floorings' service life. The results reveal the high importance of the type of protection, the type of wood, and the type of floor (with a range of estimated service life values of around 18, 17 and 16 years, respectively). This study is a first step to understanding the degradation mechanisms of the wood flooring systems, in order to extend their service life, while allowing optimising of maintenance actions, thus promoting the durability and sustainability of these floorings.

Keywords: wood floorings; degradation phenomena; service life; durability

Citation: Coelho, P.; Silva, A.; de Brito, J. How Long Can a Wood Flooring System Last? *Buildings* **2021**, *11*, 23. https://doi.org/10.3390/buildings11010023

Received: 15 December 2020
Accepted: 4 January 2021
Published: 7 January 2021

Publisher's Note: MDPI stays neutral with regard to jurisdictional claims in published maps and institutional affiliations.

1. Introduction

Nowadays, stakeholders are more aware of the relevance of healthy and sustainable buildings [1]. Recent studies [2] reveal that people in industrialized countries spend 90% of their time inside buildings and, therefore, the presence of natural elements, as wood floorings, can improve the users' physical and psychological well-being [3].

Wood has been used around the world, by different civilisations, for millennia [4], and it is one of the most common and attractive solutions for flooring worldwide. This widespread use is mainly due to the availability of this material, even in most inhabited regions of the world; however, not all wood species are appropriate for the various building uses [5]. Since wood is a natural material, its variable natural durability [6] makes it necessary to carefully evaluate its characteristics as a flooring to ensure that it is suitable for its intended use, presenting a durability compatible with the users' requirements and expectations. Therefore, when selecting a wood flooring system, users may ask "how long can a wood flooring system last?"

Romagnoli et al. [7] state that a major goal of wood technology research is still developing long-lasting wood elements with adequate mechanical performance, in other words, durable solutions. For that purpose, the wood-deteriorating agents and mechanisms that affect the wood flooring's long-term performance and service life should be identified [6]. Based on this knowledge, different degradation models can be established, considering the probable incidence of the degradation agents over time. Fundamentally, the degradation of wood elements occurs due to the presence of different degradation agents, which are usually divided into abiotic and biotic. The abiotic agents are related with weathering phenomenon, due to the exposure to damp, UV radiation, temperature, or other factors related to dynamic impacts, and abrasion and wear [8,9]. These factors are connected to physiological conditions for the presence of biotic degradation agents (e.g., subterranean insects and other xylophage insects). Both biotic and abiotic agents contribute to the deterioration of wood elements, in a complex and interconnected phenomenon [10], leading to the physical, chemical, mechanical, or biological alteration of the elements over time.

Why does knowing the service life of wood flooring systems matter? There are several important reasons for minding the wood floorings' service life. This knowledge allows adopting durability design procedures and adequate solutions at the design stage, selecting the most suitable wood type and protection for a specific use. Moreover, it allows optimising the maintenance strategies, in order to reduce the number of replacements (capitalising the investment in this flooring solution), while adopting correct measures for maintain the floors in acceptable conditions during their life cycle.

How can service life prediction be performed? The definition of reliable models for the durability and service life of wood floorings has not received as much consideration as for other materials [11]. Most of the existing studies assume an expected service life for wood floorings, but is this value credible? In reality, the actual service life of a given wood flooring depends on various factors including: (i) the quality of wood and the protection treatment; (ii) the typology and thickness of the wood elements (a parquet of solid wood can be more durable than a floor with only a few millimetres of noble wood); (iii) the conditions to which the flooring is exposed; and (iv) how well the flooring is maintained during use, besides subjective criteria. In practice, long-term data are not available, and the predictions are essentially based on manufacturers' data, accelerated testing, or extrapolations from the performance of similar materials [12]. Furthermore, current methods result from oversimplifications, and do not provide any information regarding the degradation mechanisms and the influence of critical factors that affect the performance of the wood floorings over time [13].

In this sense, in this study, a methodology for the service life prediction of indoor wood flooring is proposed, analysing in-service performance data, in order to acquire some knowledge related to the impact of the floors' characteristics in their expected service life. This research follows the methodology proposed by Gaspar and de Brito [14,15] for the service life prediction of buildings' envelope elements, which has been applied to several external façade claddings [16,17]. As a research hypothesis, it is considered that this general methodology for service life prediction of the elements of the building envelope can also be applied to predict the service life of wood flooring systems. Therefore, in this study, this methodology is applied, for the first time in the literature, to predict the service life of an element in the interior of the building. The methodology adopted is based on data collected by visual inspection, during a fieldwork survey, concerning the degradation condition of wood flooring systems. In this study, 34 Portuguese in-use houses have been inspected, corresponding to a total of 96 wood flooring systems. These data are analysed and converted into a numerical indicator, which portrays the overall degradation condition of the wood floorings analysed, which allows a graphical description of the loss of performance of these components over time and according to their characteristics. These analyses allow identifying an estimated service life of the wood flooring systems analysed, identifying the most relevant parameters for the degradation of these elements over their service life.

2. Materials and Methods

A fieldwork survey is used to collect all the relevant information for the classification of the anomalies present in the cases studies analysed, to define a model for predicting the service life of wood floorings. This survey is crucial for the definition of degradation curves, which allow evaluating the loss of performance of the wood flooring systems over time. For that purpose, 96 wood floorings, with different ages (i.e., at different stages of their service life) and various characteristics, were analysed, to obtain a degradation pattern of these floorings, through the comparison of the degradation condition of different examples. Some floors were analysed several years after installation, which makes it impossible to know in detail some constructive characteristics of the wood floors analysed, such as the thickness of the wood elements, the type of fastening to the support and the species of the wood, which can be impossible to obtain without a significant margin of error in several case studies.

The application of additional diagnosis techniques (e.g., application of laboratory tests to evaluate more specific characteristics of the wood species) may lead to more accurate results, but they are of little use in practice, both due to time constraints and additional costs for the analysis of the floors' degradation conditions. The existence of simple tools makes it possible to provide an estimation of the service life of wood flooring systems, which is certainly more relevant than making informed assumptions, due to the lack of technical or material capacity to apply more advanced or laboratory techniques for analysing wood floors. This study thus intends to provide an empirical tool, based on visual inspections, which is, in most situations, the technique used to assess the condition of the floors, assisting the decision to carry out maintenance actions, to provide some information regarding the degradation of the floors, their service life, and the factors that condition it.

The sample analysed is in Lisbon, Portugal, with construction periods between 1930 and 2018. The indoor wood floors analysed are subjected to regular maintenance, for example, either dry cleaning or wet cleaning, and, sometimes, the finishing layer is replaced. However, in the sample analysed, six case studies have been subjected to a generalised maintenance intervention (with the replacement of wood elements), and, in this scenario, it is assumed that this intervention restores the initial condition of the floor, which implies that the "age" of these floors correspond to the period between the last intervention and the inspection time.

The degradation condition of the indoor wood flooring systems is evaluated though in situ inspections, aided using a moisture-meter in singular cases of humidity problems, as anomalies caused by the presence of humidity do not necessarily imply the detection of high levels of humidity, at the time of inspection.

The anomalies observed in wood floorings can be considered a symbol of these degradation factors, which may be caused by natural or human actions, inadequate use, lack of maintenance, or unpredictable events (accidents or vandalism). Figure 1 shows some examples of the anomalies observed in the wood floorings analysed during the fieldwork survey.

In this study, the main anomalies that can occur in wood flooring systems are divided into three main groups [18,19]:

- Aesthetic anomalies, related to the visual or surface alteration of wooden floors, namely staining or colour change, cigarette marks, scratches or wrinkles, wear or detachment of the finishing layer, stains; anomalies due to inadequate maintenance, and wear of the wood material;
- Functional anomalies, which compromise the use of the flooring system (e.g., the presence of detached elements may jeopardize the users' safety) and can also affect the mechanical resistance and the performance of the wood flooring, namely, warping; swelling or other flatness deficiencies; either cracking of elements, joints, or both; broken or splintered elements; rot (identified either by changes in colour, texture in the finishing layer, or both); moisture stains; disaggregation; pulverulence; xylophage attack; and loss of wood elements; and
- Anomalies in joints, which are related to the deterioration of the filling material of the joints; the presence of these anomalies may promote the degradation of the wood flooring system. In this group, three anomalies are considered, namely, colour change of the filling material, detachment, or loss of the filling material of the joints and change of the joint size.

Different levels of degradation have been defined for each group of anomalies and, in some cases, for each specific anomaly. The degradation levels of each anomaly consider the type of anomaly and the number/size of its defects. Consequently, to distinguish the severity of the anomalies, percentages of their extent have been defined on the surfaces to be analysed. These percentages and the proposed degradation levels for each of the groups of anomalies are presented in Tables 1–3.

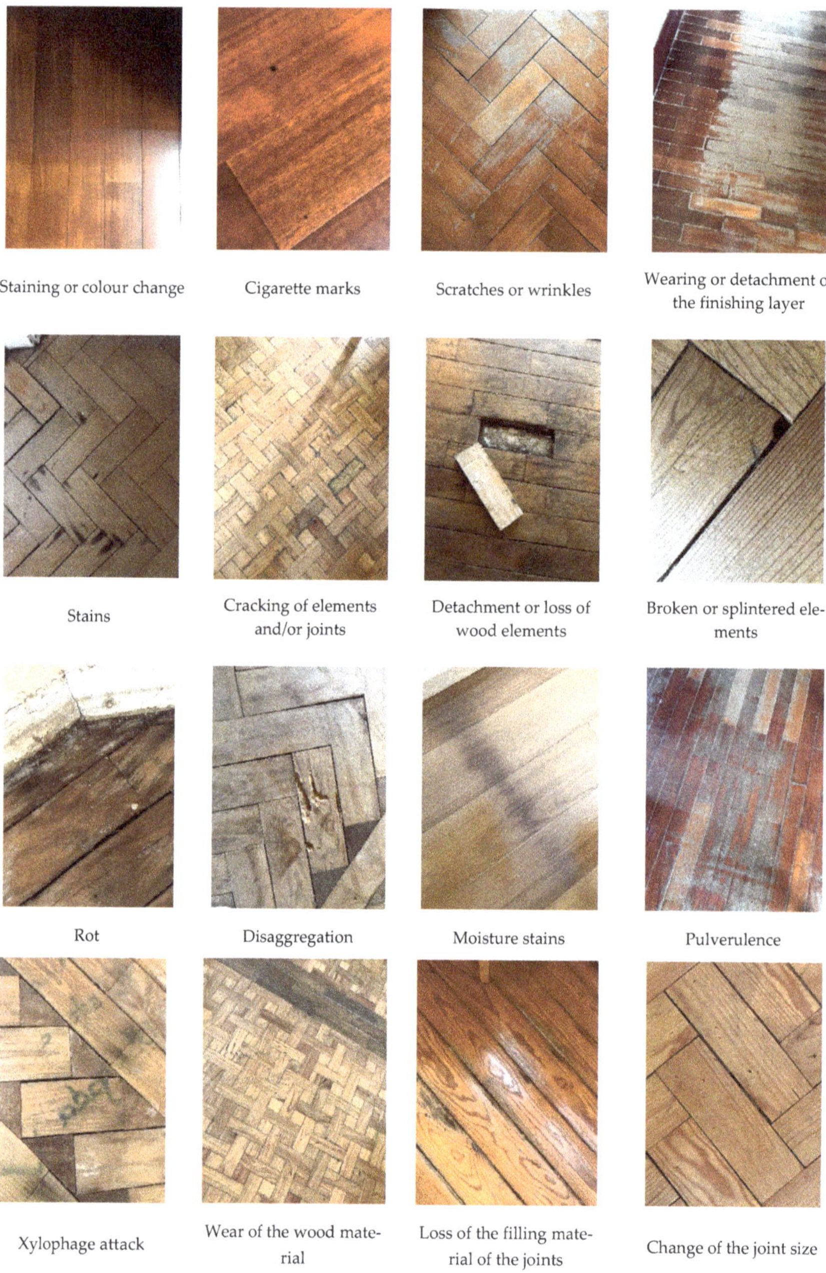

Figure 1. Illustrative examples of the anomalies observed in the wood floorings analysed.

Table 1. Degradation levels of aesthetic anomalies.

Anomalies	% Affected	Degradation Level (k_n)
Colour change	0–20	1
	20–60	2
	60–90	3
	90–100	4
Cigarette marks	0–20	3
	20–100	4
Scratches or wrinkles	0–20	1
	20–60	2
	60–90	3
	90–100	4
Wearing or detachment of the finishing layer	0–20	1
	20–60	2
	60–90	3
	90–100	4
Stains	0–10	1
	10–20	2
	20–60	3
	60–100	4
Anomalies due to inadequate maintenance	0–20	1
	20–60	2
	60–90	3
	90–100	4
Wear of wood material	0–10	1
	10–50	2
	50–90	3
	90–100	4

Table 2. Degradation level of joint anomalies.

Anomalies	% Affected	Degradation Level (k_n)
Colour change	0–20	1
	20–50	2
	50–90	3
	90–100	4
Detachment or loss of the filling material of the joints	0–20	1
	20–50	2
	50–90	3
	90–100	4
Change of joint size	0–20	1
	20–50	2
	50–90	3
	90–100	4

Table 3. Degradation levels of functional anomalies.

Anomalies	% Affected	Degradation Level (k_n)
Warping, swelling or other flatness deficiencies	0–10	1
	10–30	2
	30–90	3
	90–100	4
Cracking of elements and/or joints	0–10	1
	10–40	2
	40–90	3
	90–100	4
Broken or splintered elements	0–5	1
	5–10	2
	10–50	3
	50–100	4
Rot	0–5	1
	5–10	2
	10–50	3
	50–100	4
Moisture stains	0–5	1
	5–20	2
	20–50	3
	50–100	4
Disaggregation	0–10	1
	10–40	2
	40–90	3
	90–100	4
Pulverulence	0–10	1
	10–30	2
	30–90	3
	90–100	4
Xylophage attack	0–30	3
	30–100	4
Loss of wood elements	0–5	2
	5–20	3
	20–100	4

In this study, only the anomalies arising from use and due to the natural degradation process are considered, thus not contemplating major design or constructive errors (e.g., the presence of sapwood is considered a design characteristic and not an anomaly, in other words, it may be considered as a defect, but it is certainly a constructive defect), nor discrete events (such as vandalism actions). These phenomena are not considered, as they do not represent the natural evolution of the degradation of wood floors over time and are therefore not liable to be mathematically modelled.

These percentages were defined based on experts' opinions, considering the pathological context of the wood floors analysed, also taking into account and calibrating the degradation scale initially proposed by Prieto and Silva [18] for wooden façades. Each anomaly is rated on a discrete scale from 0 to 4, according to the following definitions: level 0 corresponds to a floor with no visible degradation; level 1 to the presence of visible anomalies, even though the floor remains in a good overall condition; level 2 to a floor that shows slight degradation signs; level 3 to a floor that shows moderate degradation; and level 4 to a floor that presents generalized degradation.

In this study, the service life prediction method initially proposed by Gaspar and de Brito [14,15] is adopted. This methodology is a deterministic empirical method, which intends to evaluate the loss of performance (or the evolution of degradation over time) of wood floorings in real service conditions and in different phases of their service life. The inclusion of the value of the areas affected by each type of anomaly in the proposed model allows assessing the extent of degradation and in parallel proceeding to the respective weighting in relation to the level of severity of each one. In this empirical method, the qualitative levels of degradation presented in Tables 1–3 are converted into quantitative information, in other words, a numerical index that establishes the overall degradation condition of the flooring, called severity of degradation (S_w). This numerical index is given by the ratio of the weighted degraded area to a reference area, equivalent to the whole flooring with the highest possible degradation level—Equation (1).

$$S_{w,wf} = \left(\sum (A_e \cdot k_n \cdot k_{a,n}) + \sum (A_f \cdot k_n \cdot k_{a,n}) + \sum (A_j \cdot k_n \cdot k_{a,n}) \right) / \left(A \cdot \sum (k_{max}) \right) \tag{1}$$

The parameters that are taken into account in the equation that is associated with the model used are: $S_{w,wf}$—severity of degradation of the wood floorings, in %; A—flooring area (m^2); A_e—area affected by aesthetic anomalies (m^2); A_f—area affected by functional anomalies (m^2); A_j—area affected by joint anomalies (m^2); k_n—multiplication factor for anomaly n, as a function of its degradation level (k varies between 0 and 4); $k_{a,n}$—weighting coefficient corresponding to the relative weight of the detected anomaly; $\sum (k_{max})$—sum of the weighing constants, corresponding to the highest possible level of degradation (4 + 4 + 4).

Therefore, after collecting fieldwork information on the degradation condition of the various wood floorings analysed, their service life is predicted using a graphical and statistical analysis of the evolution of their severity of degradation index over time.

To estimate the severity of degradation index, the weighting coefficients presented in Tables 4–6 are used. These coefficients allow obtaining values closer to reality, in other words, values that reflect, in a more adequate way, the reality observed during the inspections carried out on the floorings. The weighting coefficients are defined considering the repair costs of each anomaly, as well as the propensity of the anomaly to cause new anomalies or increase the propagation rate of the existing ones, and the effects of the anomaly in decreasing the capacity of the flooring to fulfil the minimum performance requirements.

Table 4. Weighting factors associated to functional anomalies.

Anomalies	Weighting Factor ($k_{a,n}$)
Warping, swelling, or other flatness deficiencies	1.2
Cracking of elements, joints, or both	1.2
Broken or splintered elements	1.2
Rot	1.2
Moisture	1.2
Disaggregation	1.2
Pulverulence	1.2
Xylophage attack	1.2
Crumbling	1.2

Table 5. Weighting factors associated to aesthetics anomalies.

Anomalies	Weighting Factor ($k_{a,n}$)
Colour change	0.6
Cigarette marks	0.6
Scratches or wrinkles	0.6
Wearing or detachment of the finishing layer	0.6
Stains	0.6
Improper maintenance	0.6
Wear	0.6

Table 6. Weighting factors associated to joint anomalies.

Anomalies	Weighting Factor ($k_{a,n}$)
Colour change	0.6
Detachment or loss of the filling material of the joints	1
Change of joint size	1

3. Results and Discussion

3.1. Definition of a Degradation Curve for Wood Flooring Systems

In this study, and through the adoption of the proposed model, the evolution of the degradation condition of wood floorings over time can be characterised by a degradation curve. This curve (Figure 2) is a graphical representation of the evolution of the degradation condition of wood floorings over time, and is obtained through a regression analysis between the numerical index that expresses the degradation of wood floorings ($S_{w,wf}$) and their age, considering the sample analysed in the fieldwork survey. The regression model is used to evaluate the variability of y (i.e., the severity of degradation index) that varies with the variability of x (i.e., the age of the wood floors) [16], or in other words, the regression analysis allows identifying the percentage of the variability of the severity of degradation of wood floors that are explained by their age.

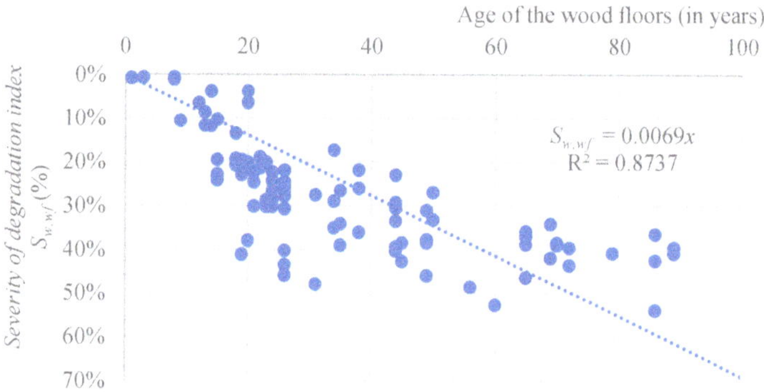

Figure 2. Regression analysis between the numerical index that expresses the degradation of the 96 wood flooring systems ($S_{w,wf}$) and their age.

In this study, a linear degradation pattern is adopted to describe the degradation of the sample analysed (97 wood floorings). According to various authors [20–23], a linear degradation pattern is adequate and valid to describe the degradation of interior components, in standard service conditions. This pattern describes a time-dependent linear degradation of wood floorings, which are subjected to consistent and continuous degradation agents over their service life, especially, when the floorings' users are the same throughout their lifetime, adopting consistent conditions of use and maintenance. Moreover, the climatic degradation agents (e.g., UV radiation, humidity, and temperature) acting on the interior floors are less aggressive and less variable (e.g., interior floors are not subjected to rainfall and direct solar radiation, whose intensity varies over the year, as well as during the day) than the agents acting on the external claddings [24].

The estimated service life of the wood flooring systems can thus be determined based on this overall degradation curve, through the intersection between the degradation curve and the theoretical limit adopted to establish the end of service life of this component. As mentioned in various studies [16,25], the end of service life is a conventional limit that it is not easy to specify, which relies on subjective criteria as the users demands and expectations, the funds available for maintenance actions, among other parameters, are difficult to model.

Figure 3 shows different examples of three levels of severity of degradation of wood floorings, revealing that a flooring with a severity of degradation of 20% still presents an adequate condition level, and therefore, adopting a maximum severity of degradation of 20% seems excessively conservative to establish the end of service life of wood flooring systems. Conversely, a floor with a severity of degradation of 40% already shows clear signs of generalised deterioration, and the presence of some anomalies that compromise the floors' functionality. In this sense, a limit of 40% for the severity of degradation, to establish the end of service life, seems too high and inadequate to fulfil the users' demands. Therefore, in this study, it is assumed that a severity of degradation of 30% corresponds to the end of service life of a wood flooring.

Figure 3. Illustrative examples of the overall condition of the wood floorings for different levels of the severity of degradation index.

Having established this limit, the estimated service life of the wood flooring systems can be obtained graphically; for the sample analysed and based on the overall degradation curve defined in Figure 2, an estimated service life of 44 years is obtained, which is in accordance with the literature and empirical knowledge about the durability of these elements. According to the existing literature on the durability of wood flooring systems, the service life of these floors can vary significantly. Nebel et al. [26] obtained an estimated service life of 10 years for multilayer parquet, while Seiders et al. [27] state that a floor can last up to 100 years, with a service life similar to the building's. Anderson et al. [28] obtained an estimated service life of 20 years for multilayer parquet glued to the substrate. Nebel et al. [26], for 8 mm parquet and 10 mm parquet, obtained an estimated service life of 25 years. Jönsson et al. [29] and Jönsson [30] estimated a service life of 40 years for pine flooring. Scharai-Rad and Welling [31] and Petersen and Solberg [32] suggested a service life of 45 years for oak flooring, while Eaton and Hale [33] and Asdrubali et al. [34] state that oak woods present an estimated service life ranging between 35 and 50 years. Adalberth [35] and Mithraratne and Vale [36] obtained an estimated service life of 50 years, for 22 mm parquet. Gunther and Langowski [37] obtained an estimated service life of more than 50 years for parquet. Aktas and Bilec [38] concluded, with an 80% confidence interval, that the service life of a wooden housing floor is, on average, 40 years. This confidence interval was defined in accordance with ISO 15686-1 [39], to define minimum and maximum limits for the variance of the results of the estimated service life, with a minimum of 15 years and a maximum of 73 years.

3.2. Influence of the Characteristics of Wood Flooring Systems on Their Service Life

The dispersion of values proposed in the literature for the expected service life of wood flooring systems reveals that the wide range of characteristics of wood floorings strongly influences their behaviour over time, affecting their durability. Reinprecht [40] refers that the service life of wood elements depends essentially on the natural durability of the wood, but also varies considerably according to the design characteristics, the protection applied, and the exposure and maintenance conditions. In this sense, the evolution of the degradation condition of wood floorings should also be analysed as a function of the different characteristics of these floors.

In this study, different curves are proposed according to the relevant characteristics of the wood floorings analysed. The type of wood is the first characteristic analysed, and five categories are considered to typify the sample analysed: (i) oak; (ii) eucalyptus; (iii) mahogany; (iv) pine; and (v) tropical woods (e.g., *Couratari oblongifolia*, *Dipteryx odorata* and *Diplotropis* sp.). In the sample analysed (Figure 4), oak and eucalyptus wood floorings present the lower estimated service lives, around 33 years, followed by mahogany floors, with an estimated service life of 35 years, and by tropical woods, with an estimated service life of 43 years, and, finally, by pine wood floorings, with an estimated service life of 50 years.

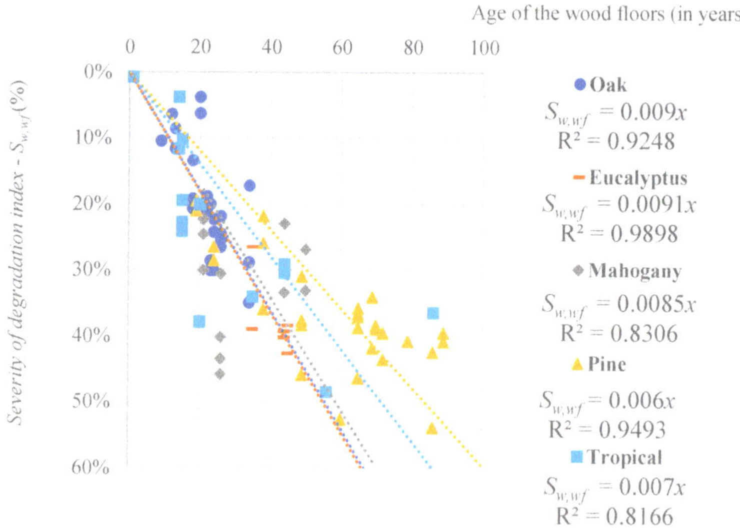

Figure 4. Regression analysis between the numerical index that expresses the degradation of the 96 wood flooring systems ($S_{w,wf}$) and their age, according to the type of wood.

Various studies [41,42] emphasise the crucial role of the natural durability of the wood species on the service life of a wood flooring. The results obtained may seem incongruous with the empirical perception regarding the durability of wood species. Usually, pine is considered as a less durable type of wood and non-resistant to decay due to xylophage attack [43,44]. However, as a natural material, the physical properties of a wood element vary significantly, even within the same wood species [45]. Cruz et al. [46] refer that, usually, the wood species are divided into softwood (e.g., pine) and hardwood (e.g., oak, eucalyptus, and some tropical species) and among hardwood species, the properties of wood range between not very resistant nor durable to very resistant and durable woods (as some Brazilian species, as the ones included in the category "tropical" analysed in the present study).

Therefore, the results reveal three major conclusions (the sample analysed is relatively small, so the results should be analysed with some caution): (i) oak, eucalyptus and mahogany species present similar degradation curves, revealing lower estimated service lives than tropical or pine wood floorings; (ii) tropical floors are usually more durable, since more naturally durable wood species have been applied while greater care was taken at the time of execution given the specificity of the material, and greater attention is taken in the floors' maintenance; and, (iii) the main reason why pine floors are more durable, within the sample analysed, is because 75% of the pine floors inspected are traditional tiles, which have a significant thickness, and whose maintenance implies polishing and applying wax, as a protective layer, weekly or monthly, thus mitigating anomalies that may occur over the service life of the floor.

In this sense, while the quality of the wood used is undoubtedly relevant to the service life of wood flooring systems, correct design also plays an important role [34]. In the sample analysed, three types of floor solutions are analysed [47–49]: (i) parquet, consisting of a set of wooden slats, with a total thickness around 8 mm; (ii) traditional floorboard, entailing several wooden strips that can be solid or laminated, with a thickness between 18 and 20 mm; and (iii) traditional tiles of solid wood, with a thickness between 17 and 22 mm. The degradation curve obtained according to the type of floor (Figure 5) reveal that parquet floors are the less durable of the sample analysed, with an estimated service life of 33 years, followed by traditional floorboard, with an estimated service life of 37 years, and by traditional tiles, with an estimated service life of 49 years. The thickness of the wood elements seems to play an important role in the durability of the wood floorings; naturally, thicker floorings, have longer estimated service lives, in addition to allowing more intrusive maintenance actions, which allow extending the service life of the elements in more adequate conditions of use.

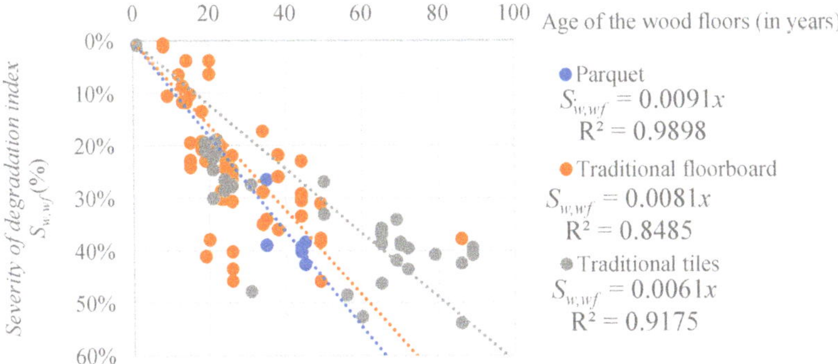

Figure 5. Regression analysis between the numerical index that expresses the degradation of the 96 wood flooring systems ($S_{w,wf}$) and their age, according to the type of flooring system.

Other intrinsic characteristics of wood may also be analysed, as its hardness, which is a relevant property for the selection of a given wood used in flooring indoors [50,51]. The hardness of the wood elements is intrinsically related with their density, which depends on several factors, such as the tree species and growing conditions, which are variable in the same species and in the same tree [3]. In this sense, the sample analysed was divided into two categories: (i) softwood species, which usually present lower densities [3]; and (ii) hardwoods, usually more resistant to impacts, scratches, and wear [52,53]. Due to the multiplicity of characteristics that influence the hardness and density of a wood species, it is not possible to draw unequivocal conclusions regarding the impact of this characteristic in the service life of wood flooring, but Figure 6 presents some statistical analysis regarding the influence of the wood's hardness in the presence of related anomalies, revealing that hardwood floorings are less prone to suffer scratches, wrinkles, and wear of the wood material.

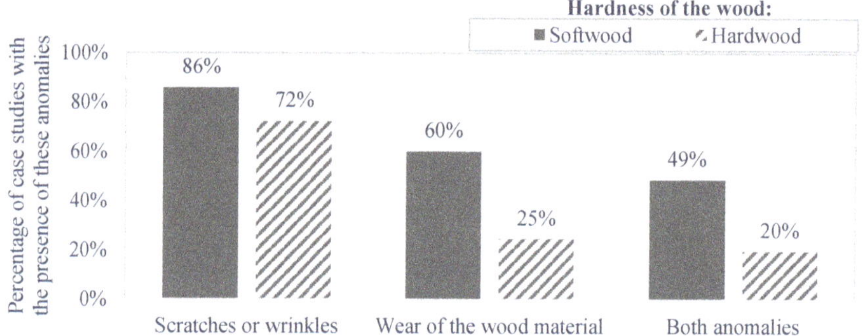

Figure 6. Statistical analysis of the presence of scratches or wrinkles and wear of the wood material, according to the hardness of the wood.

The type of surface coating of the wood floorings is also a relevant parameter to protect the floors from liquid water and direct UV radiation, avoiding anomalies that compromise the natural and aesthetic characteristics of the wood floorings, while increasing the service life of these floors [54]. Concerning the type of coating of the wood floorings, three types of protection treatments are analysed: (i) oil-based matte finishes; (ii) wax; and (iii) varnish. Figure 7 shows the degradation curves obtained for the sample under analysis, according to the surface coating. The sample corresponding to the oil-based matte finishes is small (7 case studies) and the older case study has 26 years, and therefore, no unequivocal conclusions can be drawn regarding the influence of this coating on the service life of wood floorings. The varnish coatings present a lower estimated service life (around 38 years), when compared with wood floorings with a wax coating (with an estimated service life around 52 years).

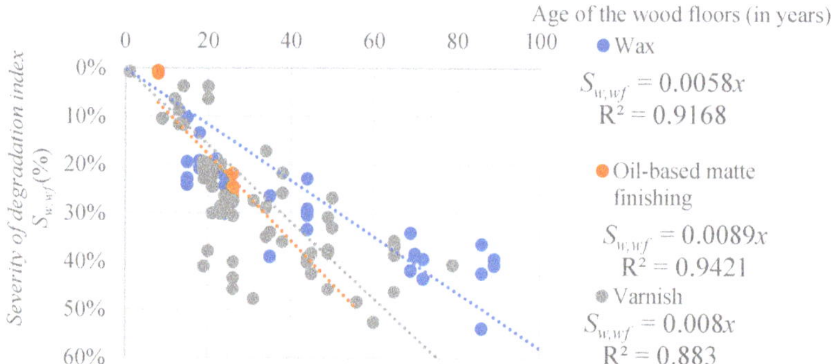

Figure 7. Regression analysis between the numerical index that expresses the degradation of the 96 wood flooring systems ($S_{w,wf}$) and their age, according to the surface coating.

Richardson [55] states that varnish can be an attractive solution to protect wood applied in interiors, but almost all varnishes suffer from a preferential wetting failure due to the fact that varnish is hydrophobic, and wood is hydrophilic, and in the presence of water, the interaction between the two materials may privilege the deterioration of the wood elements. Moreover, when exposed to the direct incidence of UV radiation, varnishes tend to oxidize and become opaque. Specifically, varnish degrades over time, and its efficiency is limited in time, losing elasticity, and becoming brittle, thus making the wood susceptible to various agents of deterioration. Unlike varnish, which is usually applied as a protective film, wax is usually impregnated in the wood [56], with generally longer periods of action, increasing the dimensional stability of the wood floorings exposed to moisture [57], and promoting a more homogeneous and regular degradation of the wood floorings over time. Furthermore, in Portugal, wood floorings with wax coatings tend to be maintained more regularly, with the removal of old wax and application of a new layer of wax.

The installation of wood floorings in contact with moisture or near moisture sources is seen as a risk factor that promotes the deterioration of these flooring systems [19,58]. Various studies [18,59,60] identify the presence of moisture as one of the main causes of anomalies in wood used as coating material. Figure 8 presents the degradation curve of wood floorings according to the exposure to moisture sources (e.g., sanitary facilities, kitchens). In the sample analysed, a wood flooring near to moisture sources has a 70% probability of presenting some anomaly related to the presence of water (e.g., moisture stains, rot, cracking). The results obtained confirm the that a wood flooring near or exposed to moisture sources reaches the end of its service life sooner (after 38 years) than a flooring protected from moisture sources, which only reaches the end of its service life after 45 years.

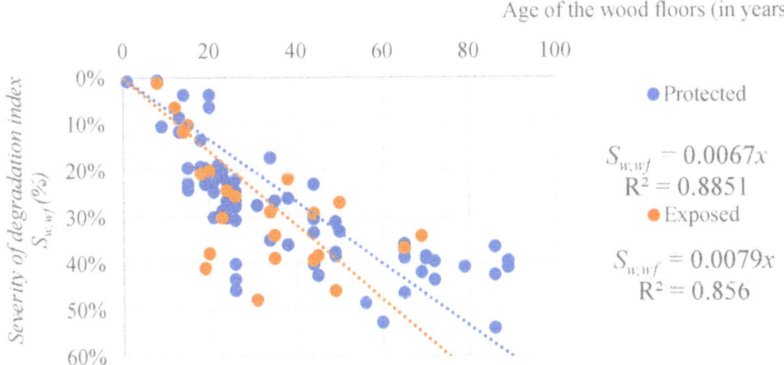

Figure 8. Regression analysis between the numerical index that expresses the degradation of the 96 wood flooring systems ($S_{w,wf}$) and their age, according to the exposure to moisture sources.

The degradation of wood floorings also occurs due to the action of mechanical or some chemical actions [46]. In the sample analysed, the case studies are not subjected to chemical degradation agents and, therefore, this degradation mechanism is not considered. On the other hand, the use conditions influence the degradation of the wood floorings analysed. Figure 9 shows two degradation curves according to the type of use and to the in-use conditions. First, the analysis of the type of use reveals that wood floorings in services buildings present a higher estimated service life (around 51 years) than floors in housing (with an estimated service life around 40 years). This result reveals that, even though the floorings in services buildings are more exposed to an intense circulation and use, a more frequent and careful maintenance increases their estimated service life.

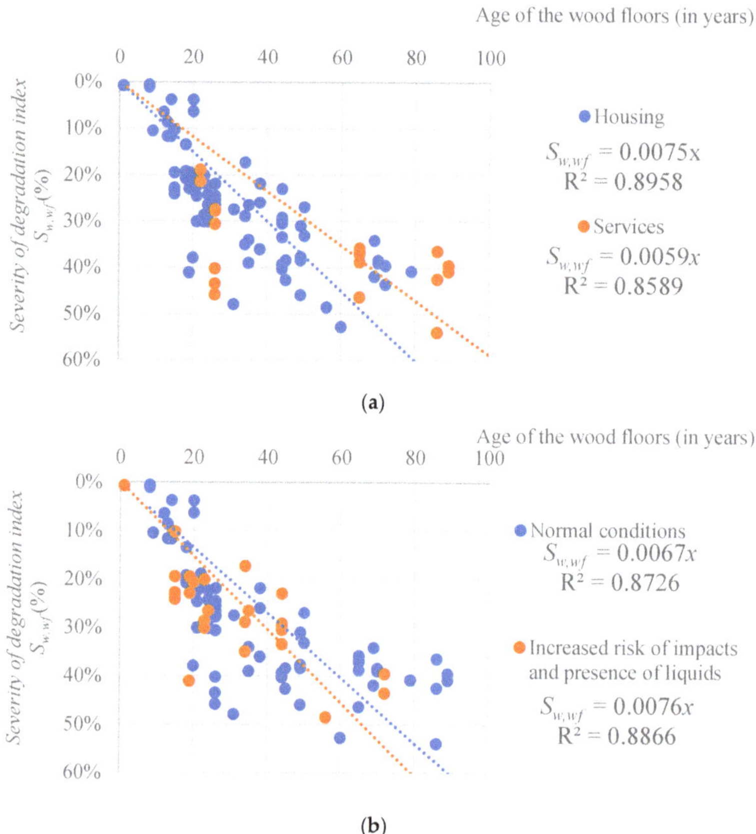

Figure 9. Regression analysis between the numerical index that expresses the degradation of the 96 wood flooring systems ($S_{w,wf}$) and their age, according to the type of use (**a**) and in-use conditions (**b**).

Moreover, in housing, some case studies are exposed to the presence of pets, children and some are used to have meals, which increases the risk of impacts and presence of liquids. When this risk is analysed, as in Figure 9b, an increased risk of impacts and presence of liquids leads to a lower estimated service life (around 39 years) than floors in normal in-use conditions (with an estimated service life of 45 years). A more detailed analysis is performed, evaluating the differences in the expected service life of wood floorings according to the house room, and although these values should be analysed with caution, given the small sample size, the following conclusions can be drawn: (i) dining rooms present the lowest estimated service life, around 28 years; (ii) followed by entrance halls, in which the circulation is more intense, with an estimated service life of 38 years; (iii) in third place come bedrooms, with an estimated service life around 42 years; (iv) classrooms and living rooms present estimated service lives of 46 and 47 years, respectively; and (v) finally, offices present an estimated service life of 49 years.

These results reveal two conflicting phenomena: on the one hand, floorings with more intense circulation should have lower estimated lives and this occurs in the case of halls. However, in the case of public buildings, offices and even classrooms, although circulation is more intense, maintenance is also more thorough and frequent, which counteracts with the higher rate of degradation of spaces. In other words, maintenance helps to preserve floors in adequate conditions for a longer time, while in residential buildings maintenance actions tend to be postponed, sometimes due to economic reasons.

Various studies [61,62] reveal the relevance of maintenance to mitigate the degradation of wood flooring systems and the impact of these actions to extend the floor's service life. In the sample analysed, the frequency of maintenance actions is divided into three categories, according to the information collected during the fieldwork survey, leading to the following results: (i) dry and wet cleaning, performed weekly, with an estimated service life of 44 years; (ii) dry cleaning, performed weekly, and wet cleaning performed monthly, with an estimated service life of 43 years; and (iii) dry and wet cleaning, performed monthly, with an estimated service life of 43 years. The differences obtained are insignificant, since all the case studies are frequently subjected to maintenance actions and, therefore, the impact of having regular maintenance cannot be fully evaluated. The type of maintenance varies between floorings, being more or less strict depending on the users; however, the actions carried out are not evaluated in detail (e.g., the cleaning products applied), thus not allowing assessing of the differences between the types of maintenance carried out in the different floorings analysed. The impact of the maintenance it is therefore more perceptible when other characteristics of the floorings are analysed, such as the surface protection or conditions of use, where the type of maintenance has an indirect impact on the floorings' service life according to these characteristics.

4. Conclusions

This study proposes an empirical degradation model to evaluate the on-site performance of wood flooring systems, considering a sample of 96 case studies. The evolution of the degradation condition of the sample analysed is discussed as a function of the different characteristics of these floors, and an estimated service was obtained for each one of the characteristics analysed. Table 7 presents a summary of the results obtained for the various characteristics of the wood floorings analysed. The sample analysed is small, from a statistical point of view; nevertheless, some statistically significant values were obtained. In this sense, a deeper statistical analysis (e.g., a multiparametric statistical analysis) would lead to debatable results, as the sample size could lead to the bias of the conclusions obtained. In future studies, with the improvement of the sample, it will be possible to carry out multiple regression analyses, which made it possible to consider the simultaneous effect of several variables on the degradation of wood floors.

The results obtained revealed, as expected, that the wide range of characteristics of wood floorings strongly influence their behaviour over time, affecting their durability. In the sample analysed, serious errors of construction or use are not considered. Additionally, when severe situations of degradation occur, the users usually perform maintenance and repair actions, in order to mitigate the progression of the degradation of the wood floorings, and, in some cases, restore their initial performance. In the sample analysed, all floorings are regularly maintained, which explains why this parameter, when grouped into the three categories considered, which already include a high maintenance periodicity, does not lead to significant differences in the overall service life of the wood floorings.

Moreover, the type of maintenance performed, as well as the cleaning products applied, are difficult to assess and can affect the state of degradation of the wood floorings. These conditions of maintenance are indirectly reflected on the use conditions. The type of use shows a variation in the estimated service life of 11 years, between floorings in public spaces and floorings in housing. The in-use conditions seem to be less relevant, with a lower impact on the overall estimated service life, with a small variation in the estimated service life (*ESL*) obtained for floorings exposed to high risks of impact and presence of liquids and floorings in normal conditions of use. The level of the users' demands associated with the sample analysed can also explain this result, since users that are more careful tend to lighten the relevance of this parameter.

The presence of moisture leads to a variation in the estimated service life of wood floorings of seven years, revealing that the proximity of a moisture source increases the incidence of anomalies related with this degradation agent, thus reducing the service life of these floorings.

Table 7. Summary of the estimated service life obtained according to the different wood floorings' characteristics.

	Characteristics	Estimated service life - ESL (years)	Determination coefficient (R^2)	Size of the sample	Impact on the overall ESL (years)
	Overall sample	44	0.874	96	-
Type of wood	Oak	33	0.925	26	−11
	Eucalyptus	33	0.99	6	−11
	Mahogany	35	0.831	13	−9
	Tropical woods	43	0.817	23	−1
	Pine	50	0.949	28	6
Type of floor	*Parquet*	33	0.99	6	−11
	Traditional floorboard	37	0.849	56	−7
	Traditional tiles	49	0.918	35	5
Type of finishing	Oil-based matte finishing	34	0.942	7	−10
	Varnish	38	0.883	60	−6
	Wax	52	0.917	30	8
Exposure to a moisture source	Exposed	38	0.856	24	−6
	Protected	45	0.885	73	1
Type of use	Housing	40	0.896	79	−4
	Public services	51	0.859	18	7
In-use conditions	Increased risk of impacts and presence of liquids	39	0.887	26	−5
	Normal conditions	45	0.873	70	1
Maintenance	Dry and wet cleaning performed monthly	43	0.921	25	−1
	Dry cleaning performed weekly, and wet cleaning performed monthly	43	0.963	9	−1
	Dry and wet cleaning performed weekly	44	0.831	62	0

Regarding the type of protection or finishing, the type of wood, and the type of flooring, these characteristics show a very significant impact on the estimated service life, in other words, a variation of 18, 17, and 16 years, respectively, is obtained for these characteristics. Therefore, these characteristics significantly influence the durability and service life of wood floorings, considering the high variation obtained for the estimated service life as a function of these characteristics, taking into account that the estimated service life of the overall sample is 44 years. This result confirms the influence of the susceptibility of wood to degradation mechanisms and the impact of design in the estimated service life of wood flooring systems.

The results obtained in this study allow quantifying of the variability introduced by the different characteristics of the wood floorings on their estimated service life, leading to an estimation in years, which is the first step to adopt more sustainable solutions during the design stage, adopting preventive measures and selecting adequate solutions for an intended use. By understanding the degradation mechanisms of the wood floorings and based on the results obtained in this study, the durability of wood floorings can be increased through the adoption of regular inspections and maintenance actions, thus promoting the sustainability of the solution and reducing its environmental impacts.

Author Contributions: Conceptualization, P.C., A.S., and J.d.B.; methodology, P.C. and A.S.; formal analysis, P.C., A.S., and J.d.B.; investigation, P.C., A.S., and J.d.B.; data curation, P.C. and A.S.;

writing—original draft preparation, P.C. and A.S.; writing—review and editing, J.d.B. All authors have read and agreed to the published version of the manuscript.

Funding: This research was funded by the Portuguese Foundation for Science and Technology (FCT) through project BestMaintenance-LowerRisks (PTDC/ECI-CON/29286/2017).

Data Availability Statement: The data presented in this study are available on request from the corresponding author. The data are not publicly available due since belongs to the research group and is part of an ongoing research.

Acknowledgments: The authors gratefully acknowledge the support of the CERIS Research Centre (Instituto Superior Técnico—University of Lisbon).

Conflicts of Interest: The authors declare no conflict of interest.

References

1. Abdul-Wahab Sabah, A. (Ed.) *Sick Building Syndrome in Public Buildings and Workplaces*; Springer: Berlin/Heidelberg, Germany, 2011.
2. Schweizer, C.; Edwards, R.D.; Bayer-Oglesby, L.; Gauderman, W.J.; Ilacqua, V.; Jantunen, M.J.; Lai, H.K.; Nieuwenhuijsen, M.; Künzli, N. Indoor time-microenvironment-activity patterns in seven regions of Europe. *J. Expo. Sci. Environ. Epidemiol.* **2007**, *17*, 170–181. [CrossRef] [PubMed]
3. Strobel, K.; Nyrud, A.Q.; Bysheim, K. Interior wood use: Linking user perceptions to physical properties. *Scand. J. For. Res.* **2017**, *32*, 798–806. [CrossRef]
4. Ramage, M.H.; Burridge, H.; Busse-Wicher, M.; Fereday, G.; Reynolds, T.; Darshil, S.U.; Wu, G.; Yu, L.; Fleming, P.; Densley-Tingley, D.; et al. The wood from the trees: The use of timber in construction. *Renew. Sustain. Energy Rev.* **2017**, *68*, 333–359. [CrossRef]
5. Larsen, K.E.; Marstein, N. *Conservation of Historic Timber Structures: An Ecological Approach*; Manual, Butterworth-Heinemann Series in Conservation and Museology; Riksantikvaren: Oslo, Norway, 2016.
6. Verbist, M.; Nunes, L.; Jones, D.; Branco, J.M. Service life design of timber structures. In *Long-Term Performance and Durability of Masonry Structures*; Ghiassi, B., Lourenco, P.B., Eds.; Woodhead Publishing: Duxford, UK, 2019; pp. 311–336.
7. Romagnoli, M.; Fragiacomo, M.; Brunori, A.; Follesa, M.; Scarascia Mugnozza, G. Solid wood and wood based composites: The challenge of sustainability looking for a short and smart supply chain. In *Digital Wood Design. Lecture Notes in Civil Engineering*; Bianconi, F., Filippucci, M., Eds.; Springer: Cham, Switzerland, 2019; Volume 24.
8. Treu, A.; Zimmer, K.; Brischke, C.; Larnøy, E.; Gobakken, L.R.; Aloui, F.; Cragg, S.M.; Flæte, P.-O.; Humar, M.; Westin, M.; et al. Durability and protection of timber structures in marine environments in Europe: A review. *Bioresources* **2019**, *14*, 10161–10184.
9. Marais, B.N.; Brischke, C.; Militz, H. Wood durability in terrestrial and aquatic environments—A review of biotic and abiotic influence factors. *Wood Mater. Sci. Eng.* **2020**. [CrossRef]
10. Björdal, C.G.; Nilsson, T. Reburial of shipwrecks in marine sediments: A long-term study on wood degradation. *J. Archaeol. Sci.* **2008**, *35*, 862–872. [CrossRef]
11. Leicester, R.H. Engineered durability for timber construction. *Prog. Struct. Eng. Mater.* **2001**, *3*, 2016–2227. [CrossRef]
12. Morrell, J.J. *Estimated Service Life of Wood Poles*; No. 16-U-101; North American Wood Pole Council: Vancouver, WA, USA, 2016.
13. Foliente, G.C.; Leicester, R.H.; Wang, C.-H.; Mackenzie, C.; Cole, I. Durability design of wood construction. *For. Prod. J.* **2002**, *52*, 10–19.
14. Gaspar, P.; de Brito, J. Service life estimation of cement-rendered facades. *Build. Res. Inf.* **2008**, *36*, 44–55. [CrossRef]
15. Gaspar, P.; de Brito, J. Limit states and service life of cement renders on façades. *J. Mater. Civ. Eng.* **2011**, *23*, 1393–1404. [CrossRef]
16. Silva, A.; de Brito, J.; Gaspar, P. *Methodologies for Service Life Prediction of Buildings: With a Focus on Façade Claddings*; Springer International Publishing: Cham, Switzerland, 2016.
17. Serralheiro, M.I.; de Brito, J.; Silva, A. Methodology for service life prediction of architectural concrete facades. *Constr. Build. Mater.* **2017**, *133*, 261–274. [CrossRef]
18. Prieto, A.J.; Silva, A. Service life prediction and environmental exposure conditions of timber claddings in South Chile. *Build. Res. Inf.* **2020**, *48*, 191–206. [CrossRef]
19. Delgado, A.; de Brito, J.; Silvestre, J.D. Inspection and diagnosis system for wood flooring. *J. Perform. Constr. Facil.* **2013**, *27*, 564–574. [CrossRef]
20. Moubray, J. *Reliability-Centred Maintenance*, 2nd ed.; Butterworth-Heinemann: Oxford, MS, USA, 1997.
21. Shohet, I.; Rosenfeld, Y.; Puterman, M.; Gilboa, E. Deterioration patterns for maintenance management—A methodological approach. In Proceedings of the Eighth International Conference on Durability of Building Materials and Components, Vancouver, BC, Canada, 30 May–3 June 1999; pp. 1666–1678.
22. Shohet, I.M.; Paciuk, M. Service life prediction of exterior cladding components under standard conditions. *Constr. Manag. Econ.* **2004**, *22*, 1081–1090. [CrossRef]
23. Lavy, S.; Shohet, I.M. A strategic integrated healthcare facility management model. *Int. J. Strateg. Prop. Manag.* **2007**, *11*, 125–142. [CrossRef]

24. Williams, R.S. Weathering of wood. In *Handbook of Wood Chemistry and Wood Composites*; CRC Press: Madison, WI, USA, 2005.
25. Moser, K. Engineering design methods for service life prediction. In *CIB W080/RILEM 175 SLM: Service Life Methodologies Prediction of Service Life for Buildings and Components, Task Group: Performance Based Methods of Service Life Prediction*; CIB: Trondheim, Norway, 2004; pp. 52–95.
26. Nebel, B.; Zimmer, B.; Wegener, G. Life cycle assessment of wood floor coverings. *Int. J. Life Cycle Assess.* **2006**, *11*, 172–182. [CrossRef]
27. Seiders, D.; Ahluwalia, G.; Melman, S.; Quint, R.; Chaluvadi, A.; Liang, M.; Silverberg, A.; Bechler, C. *Study of Life Expectancy of Home Components*; National Association of Home Builders: Washington, DC, USA, 2007; pp. 1–15.
28. Anderson, J.; Shiers, D.E.; Sinclair, M. *The Green Guide to Specification: An Environmental Profiling System for Building Materials and Components*, 3rd ed.; Consignia, Oxford Brookes University and The Building Research Establishment, Malden Blackwell Science: Oxford, UK, 2002.
29. Jönsson, Å.; Tillman, A.-M.; Svensson, T. Life cycle assessment of flooring materials: Case study. *Build. Environ.* **1997**, *32*, 245–255. [CrossRef]
30. Jönsson, A. Including the use phase in LCA of floor coverings. *Int. J. Life Cycle Assess.* **1999**, *4*, 321–328. [CrossRef]
31. Scharai-Rad, M.; Welling, J. *Environmental and Energy Balances of Wood Products and Substitutes*; Food and Agriculture Organization of the United Nations: Rome, Italy, 2002.
32. Petersen, A.K.; Solberg, B. Greenhouse gas emissions, life cycle inventory and cost-efficiency of using laminated wood instead of steel construction. Case: Beams at Gardermoen Airport. *Environ. Sci. Policy* **2002**, *5*, 169–182. [CrossRef]
33. Eaton, R.A.; Hale, M.D.C. *Wood: Decay, Pests and Protection*; Chapman and Hall Ltd.: London, UK, 1993.
34. Asdrubali, F.; Ferracuti, B.; Lombardi, L.; Guattari, C.; Evangelisti, L.; Grazieschi, G. A review of structural, thermo-physical, acoustical, and environmental properties of wooden materials for building applications. *Build. Environ.* **2017**, *114*, 307–332. [CrossRef]
35. Adalberth, K. Energy use during the life cycle of buildings: A method. *Build. Environ.* **1997**, *32*, 317–320. [CrossRef]
36. Mithraratne, N.; Vale, B. Life cycle analysis model for New Zealand houses. *Build. Environ.* **2004**, *39*, 483–492. [CrossRef]
37. Gunther, A.; Langowski, H.-C. Life cycle assessment study on resilient floor coverings. *Int. J. Life Cycle Assess.* **1997**, *2*, 73–80. [CrossRef]
38. Aktas, C.B.; Bilec, M.M. Service life prediction of residential interior finishes for life cycle assessment. *Int. J. Life Cycle Assess.* **2012**, *17*, 362–371. [CrossRef]
39. ISO 15686-1. *Buildings and Constructed Assets—Service Life Planning—Part 1: General Principles and Framework*; International Organization for Standardization: Geneva, Switzerland, 2011.
40. Reinprecht, L. *Wood Deterioration, Protection and Maintenance*; John Wiley & Sons, Ltd.: Chichester, UK, 2016.
41. Knapic, S.; Santos, J.; Santos, J.; Pereira, H. Natural durability assessment of thermo-modified young wood of eucalyptus. *MaderasCienc. Y Tecnol.* **2018**, *20*, 489–498. [CrossRef]
42. Acker, J.V.; Stevens, M.; Carey, J.; Sierra-Alvarez, R.; Militz, H.; Bayon, I.L.; Kleist, G.; Peek, R.D. Biological durability of wood in relation to end-use. Part 1. Towards a European standard for laboratory testing of the biological durability of wood. *Holz Als Roh-Und Werkst.* **2003**, *61*, 35–45. [CrossRef]
43. Thornton, J.D.; Johnson, G.C.; Nugyen, N.K. Condition of natural durability specimens from CSIRO in-ground field test after 23 years of exposure. In Proceedings of the Twenty-fourth Forest Products Research Conference, Melbourne, Australia, 15–18 November 1993.
44. Scheffer, T.C.; Morrell, J.J. *Natural Durability of Wood: A Worldwide Checklist of Species*; Research Contribution, Forest Research Laboratory publications: Corvallis, OR, USA, 1998.
45. Tenwolde, A.; McNatt, J.D.; Krahn, L. *Thermal Properties of Wood and Wood Panel Products for Use in Buildings*; National Program for Building Thermal Envelope Systems and Materials. Prepared for the U.S. Department of Energy Conservation and Renewable Energy Office of Buildings and Community Systems Building Systems Division; Oak Ridge national laboratory: Madison, WI, USA, 1988.
46. Cruz, H.; Jones, D.; Nunes, L. Wood. In *Materials for Construction and Civil Engineering*; Gonçalves, M.C., Margarido, F., Eds.; Springer: Cham, Switzerland, 2015; pp. 557–585.
47. NP 747. Floors for buildings. In *Wooden Tiles, General Characteristics and Definitions*; Instituto Português de Qualidade (Portuguese Institute of Quality): Lisbon, Portugal, 1969. (In Portuguese)
48. EN 13488. Wood flooring. In *Mosaic Parquet Elements*; British Standards Institute: London, UK, 2002.
49. Gallego, G.M. *Wood Flooring: Installation Manual*; AITIM Publishing: Madrid, Spain, 2005. (In Spanish)
50. Gurleyen, L.; Ayata, U.; Esteves, B.; Cakicier, N. Effects of heat treatment on the adhesion strength, pendulum hardness, surface roughness, color and glossiness of Scots pine laminated parquet with two different types of UV varnish application. *Maderas. Cienc. Y Tecnol.* **2017**, *19*, 213–224. [CrossRef]
51. Kokten, E.S.; Kol, H.Ş. A factorial design approach in the optimization of wood thermal modification parameters for flooring. *Pro Ligno* **2018**, *14*, 29–36.
52. Swaczyna, I.; Kędzierski, A.; Tomusiak, A.; Cichy, A.; Różańska, A.; Policińska-Serwa, A. Hardness and wear resistance tests of the wood species most frequently used in flooring panels. *Ann. Wars. Univ. Life Sci. SGGW For. Wood Technol.* **2011**, *76*, 82–87.

53. Todaro, L. Effect of steaming treatment on resistance to footprints in Turkey oak wood for flooring. *Eur. J. Wood Wood Prod.* **2012**, *70*, 209–214. [CrossRef]
54. Lozhechnikova, A.; Bellanger, H.; Michen, B.; Burgert, I.; Österberg, M. Surfactant-free carnauba wax dispersion and its use for layer-by-layer assembled protective surface coatings on wood. *Appl. Surf. Sci.* **2017**, *396*, 1273–1281. [CrossRef]
55. Richardson, B. *Defects and Deterioration in Buildings: A Practical Guide to the Science and Technology of Material Failure*, 2nd ed.; Spon Press, Taylor & Francis Group: London, UK, 2011.
56. Lesar, B.; Humar, M. Use of wax emulsions for improvement of wood durability and sorption properties. *Eur. J. Wood Wood Prod.* **2011**, *69*, 231–238. [CrossRef]
57. Kurt, R.; Krause, A.; Militz, H.; Mai, C. Hydroxymethylated resorcinol (HMR) priming agent for improved bondability of waxtreated wood. *Holz Als Roh-Und Werkst.* **2008**, *66*, 333–338. [CrossRef]
58. Garcia, J.; de Brito, J. Inspection and diagnosis of epoxy resin industrial floor coatings. *J. Mater. Civ. Eng.* **2008**, *20*, 128–136. [CrossRef]
59. Gomes, T.; Gaspar, F.; Rodrigues, H. Characterisation of building stock and its pathologies. Case study of the Historical city centre of Leiria, Portugal. In *Nondestructive Techniques for the Assessment and Preservation of Historic Structures*; Gonçalves, L.M.S., Rodrigues, H., Gaspar, F., Eds.; CRC Press, Taylor & Francis Group: Boca Raton, FL, USA, 2018.
60. Tietze, A.; Boulet, S.; Ott, S.; Winter, S. Consideration of disturbances and deficiencies in the moisture safety design of tall timber facades. In Proceedings of the World Conference on Timber Engineering, Vienna, Austria, 22–25 August 2016.
61. MacKenzie, C.; Wang, C.-H.; Leicester, R.H.; Foliente, G.; Nguyen, M. *Timber Service Life Design Guide*; Prepared for Forest and Wood Products Australia, Project No: PN07.1052; Forest and Wood Products Australia Limited: Victoria, Australia, 2007.
62. Woodard, A.C.; Milner, H.R. Sustainability of Timber and Wood in Construction. In *Sustainability of Construction Materials*, 2nd ed.; Woodhead Publishing: Amsterdam, The Netherlands, 2016.

Article

Degradation Assessment of Natural Stone Claddings over Their Service Life: Comparison between Tehran (Iran) and Lisbon (Portugal)

S. H. Mousavi [1], Ana Silva [2,*], Jorge de Brito [2], A. Ekhlassi [1] and S. B. Hosseini [1]

[1] School of Architecture & Environmental Design, Iran University of Science & Technology, Hengum Street, Resalat Square, Tehran 13114-16846, Iran; hosein_mousavi@hotmail.com (S.H.M.); ekhlassi@iust.ac.ir (A.E.); hosseini@iust.ac.ir (S.B.H.)

[2] Department of Civil Engineering, Architecture and Georesources, Instituto Superior Técnico, Universidade de Lisboa, Avenue Rovisco Pais, 1049-001 Lisbon, Portugal; jb@civil.ist.utl.pt

* Correspondence: anasilva931@msn.com

Abstract: Now more than ever, the construction sector is aiming to adopt more sustainable solutions. To achieve this purpose, more durable solutions must be adopted, making rational decisions at the design and maintenance stages regarding the conditions of environmental exposure and use. In this sense, knowledge regarding the service life of building components is crucial. This knowledge should not be a general concept, or a standard value, and adapting practices from one country to another is extremely challenging. In this sense, this study analyses the service life of natural stone claddings. We adopt a methodology initially proposed for Lisbon (Portugal), intending to evaluate its applicability to other geographical contexts, in order to perform a more reliable service life prediction of stone claddings located in Tehran (Iran). An estimated service life of 65 years was obtained for a sample of 162 stone claddings directly adhered to the substrate, located in Tehran, which were analysed by in situ inspections. The impact of different conditions (e.g., type of stone and environmental exposure conditions) on the service life of stone claddings in Tehran was quantified, which revealed that the exposure to environmental agents, such as wind, rain and pollutants, is the main cause of degradation of the natural stone claddings.

Keywords: durability; in situ inspections; service life prediction; natural stone claddings

Citation: Mousavi, S.H.; Silva, A.; de Brito, J.; Ekhlassi, A.; Hosseini, S.B. Degradation Assessment of Natural Stone Claddings over Their Service Life: Comparison between Tehran (Iran) and Lisbon (Portugal). *Buildings* **2021**, *11*, 438. https://doi.org/10.3390/buildings11100438

Academic Editor: Wahidul Biswas

Received: 27 July 2021
Accepted: 23 September 2021
Published: 27 September 2021

Publisher's Note: MDPI stays neutral with regard to jurisdictional claims in published maps and institutional affiliations.

1. Introduction

The durability of buildings is one of the fundamental elements in assessing the quality of life of contemporary societies and is indispensable to their social and economic stability [1]. The durability of buildings can be jeopardised by increasing the degradation of their components, resulting from the natural ageing process of materials [2], leading to the loss of a building's performance [3]. Due to this degradation process, the buildings tend to become obsolete and unable to meet the performance demands established during the design stage [4]. Therefore, the service life prediction of building elements is of utmost value for a more sustainable environment [5]. The knowledge about the durability of building elements allows a more rational management of resources, which can reduce the economic and environmental costs of buildings during their lifetime [6].

Building components are categorised in various studies [7–10] according to the time they take to reach the end of their service life, i.e., as durability layers. The façade is considered the "skin" of the building, acting as a protective layer against environmental degradation agents [2]. Therefore, the cladding is more prone to anomalies [11], which is why the building envelope failures can be held accountable for more than 50% of all building deficiencies [12]. The adoption of natural stone as a cladding solution is growing due to its durability and mechanical strength [13].

A review of the literature shows that a variety of approaches can be used in the description of models to replicate the degradation of façade claddings during their lifetime. Gaspar and de Brito [14] classified these under two categories, namely, (i) laboratory testing and (ii) fieldwork surveys, to evaluate the degradation condition of the elements in real exposure situations. In spite of using laboratory testing methods because of the quickness with which they yield results [15], some authors indicate that the simulation of the complete effect of environmental agents that simultaneously contribute to the degradation of the building elements is not possible in such an artificial environment, since the degradation effects are analysed in a much reduced way [16]. In addition, real life situations and the synergy between different degradation agents are difficult to model in a laboratory setting [17]. The studies conducted by Searls and Tomasen [18] and Henriksen [19] reveal that laboratory tests are more appropriate to model the effects of a given degradation agent [20].

To overcome this criticism, an evaluation of the influence of the combined actions of different agents on claddings' service life by observing the existing condition was used. This research method, called ex post facto, which literally means 'after the fact' or 'retrospectively', refers to those studies that investigate possible cause-and-effect relationships by examining the elements under analysis in real situations and searching back in time for plausible causal factors [21]. This course of action to assess the durability of building components was conducted in several studies [22–28].

However, most of the research in the field of the degradation processes of exterior claddings in real conditions in retrospect have been carried out in a specific region with a specific climate type, and the identification of the influence of some properties on the evolution of the degradation can come to the light by comparing the same building components in different climate types. In this sense, Haagenrud [3] suggests that the relation between the degradation agents and their effects on the degradation process can be explained when the different geographical coordinates and climates are taken into consideration in the characterisation and mapping of the most significant degradation agents. Therefore, the impact of environmental agents and façade specifications determined by comparing the natural stone cladding behaviour in Tehran (Iran) and Lisbon (Portugal), performed within a PhD study [29] and unique in the literature, is described in this study.

Regardless of the innovations in cladding materials used in exterior walls, natural stone is still one of the most efficient and durable options [13]. Therefore, this cladding solution was selected as a case study in the present study. For decades, many methods to determine the adherence of stone plates to a building's external walls have been developed, which can be classified under the headings of (i) a 'direct fastening system', which corresponds to the complete bond between the substrate and the stone elements, and (ii) an 'indirect fastening system', adopting mechanical anchorage methods [30,31]. Considering the fact that the indirect fastening system is a relatively new technology and that it is therefore difficult to find a significant number of case studies, natural stonewall claddings with a direct fastening to the substrate were chosen in the current study. The sample analysed comprises 203 stone claddings located in Lisbon (Portugal) and 162 stone claddings in Tehran (Iran), which were inspected in situ.

The aim of this study is essentially to propose an empirical method to establish a degradation path for stone claddings over time, through a simple tool to predict the service life of these claddings, considering characteristics that can be easily assessed by visual inspections in situ. This study intends to analyse and quantify the impact of different factors, such as the characteristics of the stone elements and environmental exposure conditions on the service life of natural stone claddings in Tehran. This knowledge is extremely useful for the adoption of more sustainable and rational solutions at the design stage, as well as for the definition and optimisation of maintenance strategies to promote the durability of these claddings.

2. Materials and Methods

2.1. Climatic and Environmental Exposure Conditions

In this study, according to Haagenrud [3], different geographical coordinates and climates are taken into consideration at various scales. The macro scale usually describes the overall meteorological conditions based on meteorological indicators (e.g., air temperature, precipitation, among others). Tehran is categorised as having a cold semi-arid climate (BSk) based on the Köppen–Geiger classification [32], which denotes that at least one month's average temperature is below 0 °C. By contrast, the Lisbon area is situated in a hot summer Mediterranean climate (Csa), where the average temperature is above 0 °C even in the coldest months. With respect to the *meso* climate, the effects of the ground and of the built environment are considered. In this context, Tehran is spread from 35°34′ to 35°51′ north latitude and from longitude 51°06′ to 51°38′ east with an average elevation of 1190 m, comprising 22 regions. The current study focused on the sixth district, which is geometrically located in the centre of Tehran and was founded in the 1950s (Figure 1).

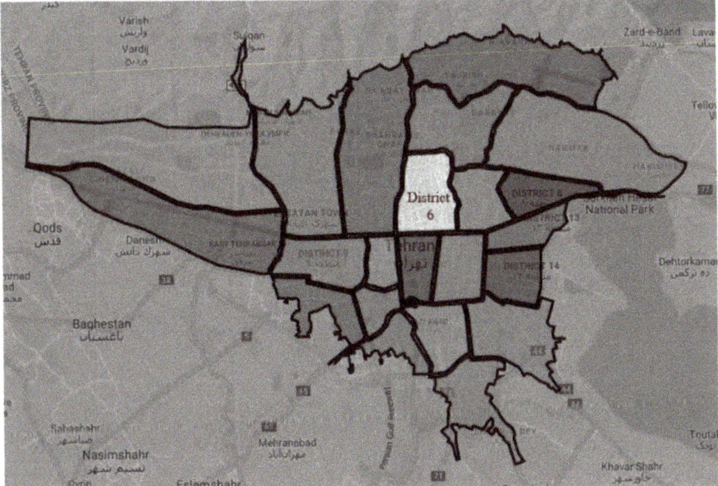

Figure 1. Map of the 6th district, located in the centre of Tehran.

The other research area is in the Lisbon area at 42°38′ north latitude and 09°08′ west longitude, located in western Portugal on the estuary of the Tagus River along the Atlantic coast.

The impact of environmental conditions on cladding degradation varies according to the building's exposure circumstances, which mainly depends on the local climate [33,34]. Accordingly, the local specifications, i.e., the local conditions in the building's proximity, were collected at the data acquisition stage.

2.2. Data Acquisition and Limitations of the Method Proposed

As Silva et al. [2] mentioned, in order to appraise the service life of building elements considering their current condition, various methods can be used, including destructive and non-destructive tests carried out in situ. The former procedures, however accurate in regarding the information provided, usually demand costly equipment, in addition to requiring repair actions afterwards. By contrast, the latter techniques, notably visual inspections, are generally fast, less expensive and adequate to establish the degradation condition of the elements under analysis [35]. Moreover, in most real situations in the geographical contexts analysed, the maintenance actions are carried out based on subjective criteria that depend on the users' perception of the state of degradation of the façades acquired through simple visual inspections.

Consequently, in the current study, an in situ survey of buildings, through visual inspections, was used to collect data regarding the degradation condition of natural stone claddings. The fieldwork survey encompasses the following steps (Figure 2): (i) each case study is selected and characterised based on complementary information, such as the location of the case study, existing drawings and other relevant documents from the municipality; (ii) an inspection file is used to collect all the relevant information to characterise the case study (e.g., façade's dimensions, exposure conditions and the description of the anomalies observed, as well as the area of the façade affected by these anomalies); (iii) during fieldwork, the case study is photographed, and, aided by diagrammatic sketches, the anomalies observed are identified and their extent is quantified using design and calculus software.

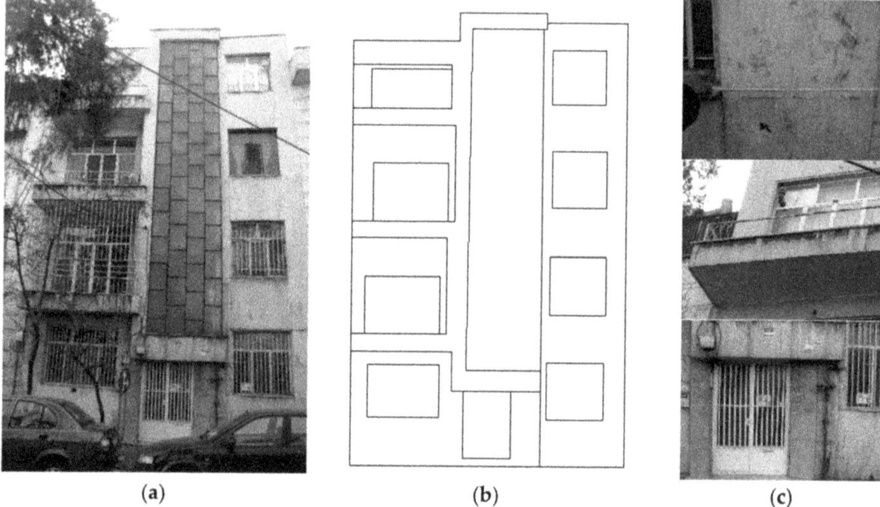

(a) (b) (c)

Figure 2. Illustrative example of one case study analysed in Tehran: (**a**) photograph of the façade; (**b**) diagrammatic sketch to represent the anomalies observed; (**c**) direct measurements and identification of different anomalies in the natural stone cladding.

The method proposed, in which the fieldwork is based on visual inspections only, presents some limitations regarding the collection of relevant information to characterise some properties of the claddings, namely (i) the factors related to the characteristics of the materials applied in the execution stage, such as the bonding material applied or the material used to fill the joints, cannot be determined during the visual inspections; (ii) the execution conditions and the level of control over the construction processes (e.g., the quality of the workforce work, material storage environments, among others) are also difficult or even impossible to characterise. The use of inadequate materials or execution errors irreparably affect the durability of stone claddings, but there is a lack of information in municipalities about the execution processes, which do not allow consideration of these factors when using visual inspections only. However, in this study, the case studies that presented anomalies due to clear execution errors or choice of inappropriate materials were excluded from the analysis, as they do not characterise the physical degradation phenomenon of natural stone claddings.

Information about the maintenance and rehabilitation actions carried out on the façades analysed was collected in the municipalities and through contact with the owners. The type of actions carried out is not within the scope of this study, as the models consider the natural evolution of degradation of stone claddings without maintenance. In other words, the "age" is considered as the period since construction/last intervention/maintenance action until the inspection time.

Despite these limitations, the method used has numerous practical advantages. The main advantage relies on the time required to obtain relevant results regarding the influence of exposure conditions on the degradation of the claddings. The combined action of the degradation agents is not easily reproduced. In this sense, the field data on the degradation condition of buildings should be used when possible, since they provide relevant information on the real degradation of components when exposed to a particular set of degradation agents, which act simultaneously and synergistically, whose action it is not possible to simulate with precision in an artificial way.

3. Description of the Samples Analysed

Accordingly, with respect to the samples located in the Lisbon area, fieldwork data from previous research [22,24] were analysed in this study, related to the inspection of 203 stone claddings. The newest building was built in 2008, and the oldest building was built in 1891 (being subjected to a maintenance intervention in 1948). As for the case studies in Tehran, based on the research of Silva et al. [22] for the classification system of anomalies in stone claddings with direct fastening to the substrate, a visual inspection plan was developed. This fieldwork generated data from 52 buildings in the sixth district of the Tehran (Iran) area, whose claddings presented various anomalies, totalling 162 case studies. In the Tehran dataset, the oldest building is from 1969 and the newest one from 2010. The resulting data can generally be arranged into the headings of "materials' characteristics" and "defects". The former category comprises the type of stone, size and colour; the type of finishing; and the position of the cladding in the façade; in addition, complementary information (e.g., orientation, exposure to wind-driven rain, among other parameters) was collected in advance to put each case study into a better perspective (Table 1).

Table 1. Description of the samples collected in Tehran and in Lisbon.

Cladding Characteristics		Tehran (162 Cases)		Lisbon (203 Cases)	
		Number of Cases	% of Cases	Number of Cases	% of Cases
Type of Stone	Granite	52	32.1%	54	26.6%
	Limestone	47	29.0%	72	35.5%
	Marble	63	38.9%	77	37.9%
Colour of natural stone	Light colours	101	62.3%	134	66.0%
	Dark colours	61	37.7%	69	34.0%
Type of finishing	Smooth	108	66.7%	96	47.3%
	Rough	54	33.3%	107	52.7%
Size of stone plates	Medium size	105	64.8%	129	63.5%
	Large size	57	35.2%	74	36.5%
Location of the cladding	Integrally or partially elevated	116	71.6%	61	30.0%
	Partial bottom	46	28.4%	142	70.0%
Orientation	South/southwest	20	12.3%	35	17.2%
	East/southeast	32	19.8%	60	29.6%
	West/northwest	32	19.8%	51	25.1%
	North/northeast	78	48.1%	57	28.1%
Height of the building	Current (≤5 floors)	122	75.3%	110	54.2%
	High (>5 floors)	40	24.7%	93	45.8%
Use of the building	Residential	149	92.0%	104	51.2%
	Mixed use	13	8.0%	99	48.8%
Exposure to wind/rain	Moderate	58	35.8%	170	83.7%
	Severe	104	64.2%	33	16.3%

According to Silva et al. [2,22], four groups of anomalies are considered:

(i). Aesthetic degradation, which has a visual impact on the cladding, not jeopardising its integrity; these anomalies can be observed in most stone claddings and tend to occur prematurely;

(ii). Joint defects, which affect the proper functioning of the joints of the cladding and can promote the occurrence of new anomalies;

(iii). Fastening to the substrate that jeopardises the claddings' and the users' safety;

(iv). Loss of integrity due to the modification of the physicochemical properties of the stone (e.g., due to the exposure to pollutants), leading to irreversible changes in the physical and aesthetic properties of the claddings.

From the preliminary analyses of the data collected, as shown in Figure 3, the most and least frequent defects detected during the fieldwork in both research areas are aesthetic anomalies and those related to the fastening to the substrate, respectively. In the Tehran dataset, the aesthetic anomalies are followed by defects related to the loss of integrity (32%) and, finally, joint defects (15%). On the contrary, the analysis of the Lisbon cases shows joint defects in 29% of the inspected façades, while the defects associated with the loss of integrity occur in 28% of the case studies analysed.

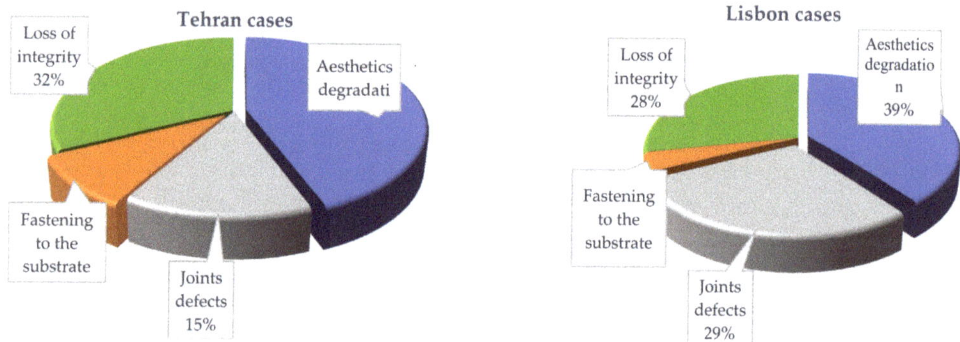

Figure 3. Statistical analyses of the defects observed in the case studies analysed.

The frequency of the different defects in the aesthetic anomalies group observed in the cases under analysis can be compared in Figure 4. Among the aesthetic degradation defects in both research areas, superficial dirt, caused by the accumulation of pollution debris on the façade, is the most common one. In Tehran, this is followed by colour change and stains (occurring in 46% and 44% of the sample analysed, respectively), while an inverse tendency in these defects is observed in the inspected façades in Lisbon. The fourth ranked frequency of the aesthetic degradation defects observed in both fieldwork surveys relates to flatness deficiencies, which is followed by efflorescence and biological colonisation, respectively, in Tehran, whilst the Lisbon samples exhibit inverse behaviour. The incidence of vegetation growth only occurs in the Lisbon sample.

Concerning the defects related to the joints, as shown in Figure 5, degradation of the materials used to fill the joints is more frequent than the loss of material (which corresponds to a more serious defect with severe consequences, such as penetration of water in the substrate). Concomitantly, given the higher age of the samples under analysis in Lisbon (less than 89 years) than in Tehran (less than 48 years), the first sample was naturally subjected for a longer period to the degradation mechanisms, thus leading to a higher incidence of joint defects.

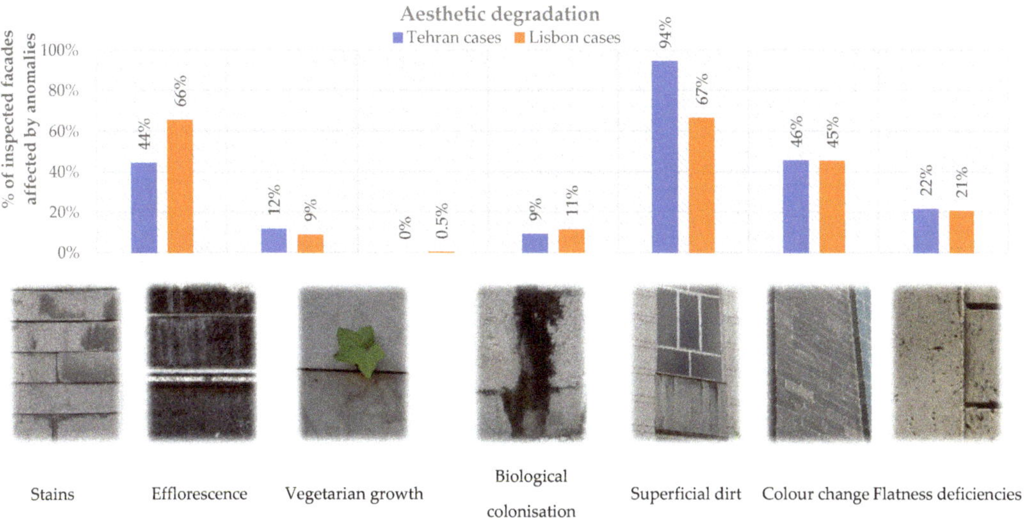

Figure 4. Frequency of the defects in the aesthetic anomalies group observed in the samples under analysis.

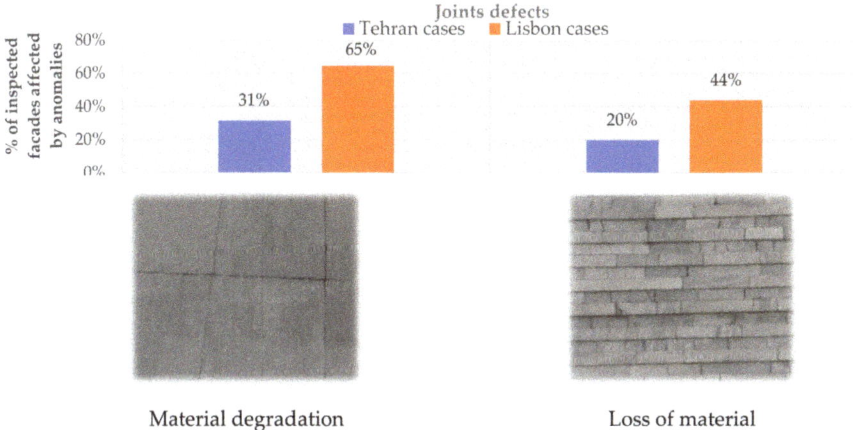

Figure 5. Frequency of the joints' defects in the samples under analysis.

Concerning the defects related to the fastening to the substrate (Figure 6), the scaling of the stone near the edges is the major defect in the sample analysed. This is followed by a partial loss of stone material and detachment, respectively, while a partial gap in the stone elements does not occur in any inspected cases in Tehran. Conversely, the analysis of the Lisbon data reveals that the presence of partial gaps in the stone elements is the second most common defect, followed by a partial loss of stone material.

Notwithstanding the lower age of the studied samples in Tehran (with an average age of 26 years and a maximum age of 48 years) in comparison with those in Lisbon (with an average age of 38 years and a maximum age of 89 years), the defects related to the loss of integrity are more frequent in Tehran than in Lisbon, as presented in Figure 7. This result can be justified by the presence of high levels of pollutants in Tehran, which occasionally reach dangerous levels, jeopardising the health of its residents and their quality of life [36,37]. Higher air pollution increases the deterioration and weathering of stone materials [38], causing different types of anomalies related to the loss of integrity of stone plates (e.g., erosion). Moreover, the action of pollutants also promotes discoloration of the stone elements [39], as observed in 46% of the Tehran sample, and a higher deposition of sediments (Figure 4).

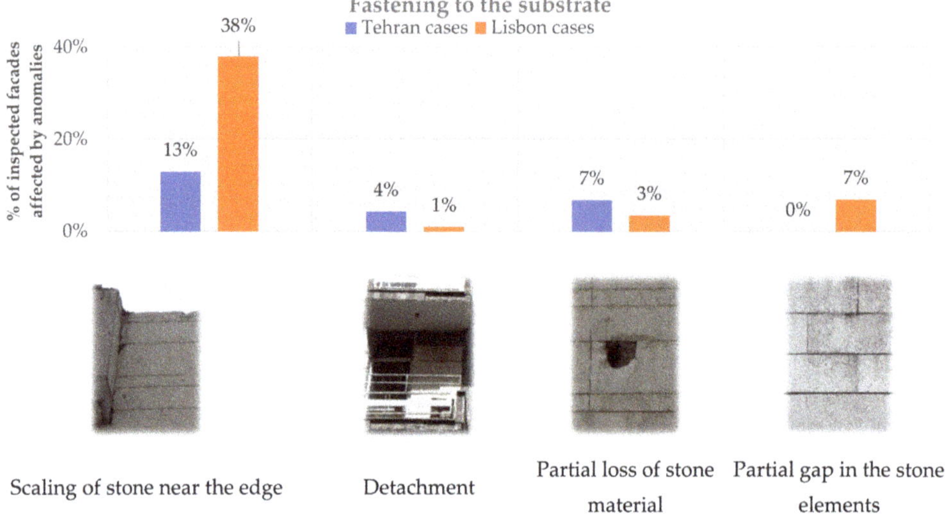

Figure 6. Frequency of the fastening to substrate defects in the samples under analysis.

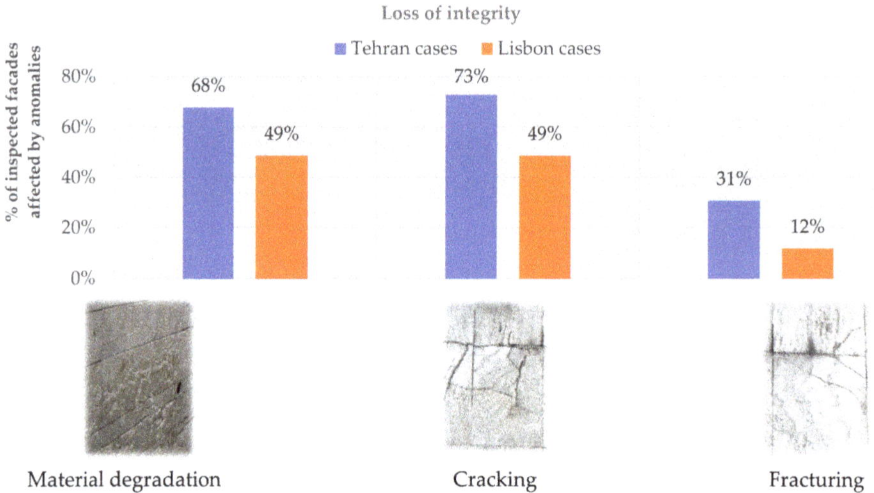

Figure 7. Frequency of the different loss of integrity defects in the samples under analysis.

In this study, the defects inspected visually during the fieldwork survey are discreetly ranked based on the proposal defined by Silva et al. [2,22] regarding the degradation condition of natural stone claddings. In this classification system, which was used in several studies [24,25,27,40], the defects are ranked from 0 (no visible degradation) to 4 (generalised degradation), considering the area of the façade affected by each anomaly and its relative importance within the degradation scale adopted (Table 2).

Table 2. Classification of the degradation condition of natural stone claddings (data sourced from Silva et al. [2]).

Degradation Level		Anomaly	% Area of Cladding Affected [**]
Level 0		*No visible degradation*	-
Level 1 *Good*	Aesthetic degradation anomalies	Surface dirt	>10%
		Moisture stains	
		Localised stains	≤15%
		Colour change	
		Flatness deficiencies	≤10%
	Loss-of-integrity anomalies	Material degradation [*] ≤ 1% plate thickness	-
		Material degradation [*] ≤ 10% plate thickness	≤20%
		Cracking width ≤ 1 mm	
Level 2 *Slight Degradation*	Aesthetic degradation anomalies	Moisture stains	>15%
		Localised stains	
		Colour change	
		Biological colonisation	
		Vegetation growth	≤30%
		Efflorescence	
		Flatness deficiencies	>10% and ≤50%
	Joint anomalies	Joint material degradation	≤30%
		Material loss—open joint	≤10%
	Fastening to the substrate anomalies	Scaling of stone near the edges	≤20%
		Partial loss of stone material	
	Loss-of-integrity anomalies	Material degradation [*] ≤ 10% plate thickness	≤20%
		Material degradation [*] > 10% and ≤ 30% plate thickness	≤20%
		Cracking width ≤ 1 mm	≤20%
		Cracking width > 1 mm and ≤ 3 mm	≤20%
		Fracture	≤5%
Level 3 *Moderate degradation*	Aesthetic degradation anomalies	Biological colonisation	>30%
		Vegetation growth	
		Efflorescence	
		Flatness deficiencies	>50%
	Joint anomalies	Joint material degradation	≤30%
		Material loss—open joint	≤10%
	Fastening to the substrate anomalies	Scaling of stone near the edges	>20%
		Partial loss of stone material	
		Detachment	≤10%
	Loss-of-integrity anomalies	Material degradation [*] > 10% e ≤ 30% plate thickness	>20%
		Material degradation [*] ≤ 30% plate thickness	≤20%
		Cracking width > 1 mm and ≤ 3 mm	>20%
		Cracking width ≥ 3 mm	≤20%
		Fracture	>5% and ≤10%
Level 4 *Generalised degradation*	Fastening to the substrate anomalies	Detachment	>10%
	Loss-of-integrity anomalies	Material degradation [*] > 30% plate thickness	>20%
		Cracking width > 3 mm	
		Fracture	>10%

[*] Material degradation is meant to be every anomaly that involves loss of volume of the stone material. [**] Data from Silva et al. [2,22].

A preliminary analysis of the Tehran data shows that every cladding presents visible defects. Most of the anomalies observed are ranked level 1 (52% of the sample), followed by level 2 (45%) and, finally, by level 3 (3%); none of the inspected claddings belong to level 4. In the Lisbon data, the defects ranked in level 2 are the most common with a frequency of

55%, followed by those in levels 1 and 3, which occur in 27% and 16% of the case studies, respectively, and, finally, anomalies related to level 4 occurred in only 2% of the façades in question [2,22].

4. Service Life Prediction Model

According to the literature, several methods can be used to predict the service life of claddings over time, which can be categorised as analytical, statistical, empirical and experimental [41]. Regarding the probabilistic nature of degradation phenomena in addition to the capability of statistical models to establish probabilistic and statistical means to predict the performance and failure probability [42], the statistical model proposed by Gaspar and de Brito [14] was adopted in the current study. This model was effectively used for the quantification of the degradation condition of various types of claddings and in other geographical contexts, namely in South America [43,44], and this is the first application of this model in Asia. This approach relies on the calculation of a numerical index (severity of degradation), which is given by the ratio among the area of the façade affected by the different anomalies (weighted based on the degradation level and the severity of the anomalies observed), and a reference area corresponding to the maximum hypothetical extent of the degradation for the façade under analysis—Equation (1).

$$S_w = \frac{\Sigma(A_n \times k_n \times k_{a,n})}{A \times \Sigma(k_{\max.})} \tag{1}$$

where S_w—normalised severity of degradation of the façade as a percentage; A_n—area of the cladding affected by a defect n, in m^2; k_n—defects' "n" multiplying factor as a function of its condition (between 1 and 4); $k_{a,n}$—weighting coefficient to encompass the relative consequences of each defect, which considers the cost of repair of defect n ($k_{a,n} \in R^+$) (in absence of more information, $k_{a,n} = 1$); k_{max}—weighting factor equal to the worst condition level; and A—total area of the cladding, in m^2.

In this study, the groups of defects are weighted according to their repair costs [22] (i.e., the ratio between the total cost of the intervention needed to repair the anomaly and the cost of replacing the cladding). A higher weighting coefficient is thus assigned to the more complex and costly anomalies as shown in Table 3. The values of the weighting coefficient ($k_{a,n}$) are different for the two samples, since these values are adjusted to the costs applied in each geographical context.

Table 3. Weighting coefficients corresponding to the relative importance of each group of defects.

Anomaly	Repair Operation	Weighting Coefficient ($k_{a,n}$)	
		Tehran	Lisbon [*]
Aesthetic degradation	Cleaning	0.12	0.13
Joints	Joint repair	0.13	0.25
	Replacement of the joint materials	1.00	1.00
Fastening to the Substrate	Replacement of stone plates	1.23	1.20
Loss of Integrity	Repairing the affected stone plates	1.02	1.00

[*] Data from Silva et al. [22].

To evaluate the degradation of stone claddings over time, four typical deterioration patterns were proposed by Shohet et al. [20], defined based on specific degradation agents and mechanisms, comprising linear, convex-shaped, concave-shaped and S-shaped patterns. In this study, an S-shaped pattern was used, which allows taking into account different rates of the degradation of stone claddings; i.e., it is capable of describing the process of the cladding's deterioration, which starts slowly and apparently stabilises over time but accelerates again near the end of the cladding's service life. Accordingly, "S"-shaped degradation curves are obtained through a nonlinear regression, where a third-degree polynomial line is fitted to the case studies analysed in the fieldwork (Figure 8). The degradation curves of the Tehran and Lisbon data sets lead to a determination coefficient (R^2) of 0.80 and 0.77, respectively. This reveals that a high percentage of the variance of the variable "severity" can be described by the proposed curves, i.e., there is a good correlation between the observed values and those predicted by the model.

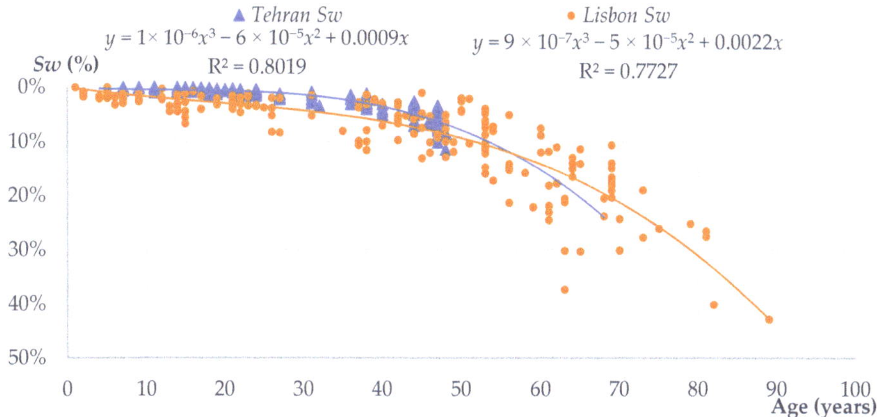

Figure 8. Degradation curves of the Tehran and Lisbon samples.

To estimate the service life of the natural stone claddings, a threshold degradation level that establishes the end of its service life must be defined. The relevant literature on this subject suggests that establishing the conventional limit for the end of service life is not simple, depending on acceptance criteria, not certainly determined by "pure" scientific methods. Therefore, and according to previous research on natural stone claddings [2,22,27], a degradation severity level of 20% was chosen in this study to establish the end of service life. Accordingly, an estimated service life of 65 years was obtained, using the degradation curves in Figure 8, for the claddings located in Tehran, and an estimated service life of 68 years was calculated for the claddings in Lisbon.

5. Analysis of the Degradation Evolution According to Claddings' Characteristics

Various studies [2,10] discuss the relevance of several contributory factors for the presence of defects in natural stone claddings, which are responsible for the claddings' deterioration. In fact, the stone claddings show a distinct behaviour in terms of deterioration because of the great variability of their characteristics. In order to compare the degradation evolution in Tehran and Lisbon, the claddings' features chosen in the research conducted by Silva et al. [2] for the Lisbon samples were also analysed in this work for the Tehran cases. They comprise the type of stone and related properties (colour, size and type of finishing), location of the cladding in the facade, the facade's orientation and their exposure to environmental conditions (wind/rain action).

Natural stone is generally classified into three groups—igneous, sedimentary and metamorphic [30,31]—according to its mineral and chemical composition, the texture of the constituent particles and the genesis processes. Therefore, the samples under analysis here were grouped into three categories: (i) granite and similar stones, (ii) travertine and limestone and (iii) marble and crystal. Figure 9 provides the degradation evolution over time according to the type of stone plates used in the claddings.

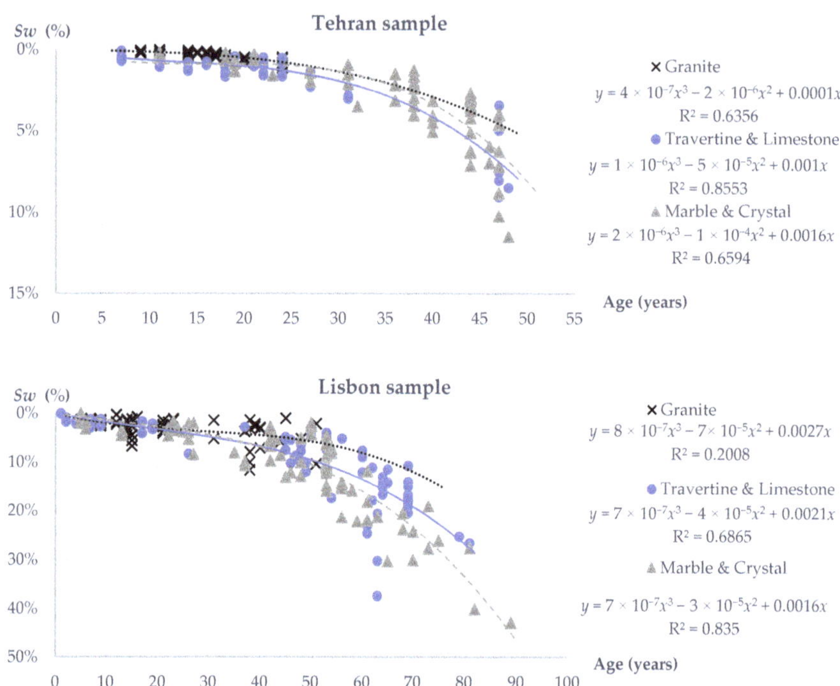

Figure 9. Degradation curves of the Tehran and Lisbon samples, according to the type of stone.

In both research areas, granite stones present a longer estimated service life, even though the Lisbon data show a curve that is statistically unreliable due to a high scatter of results. In second and third place, the degradation curves reveal that travertine and limestone are more durable than marble and crystal in exterior wall claddings. An estimated service life of 64 and 63 years was obtained for limestone and marble, respectively, in Tehran, while 78 and 66 years were the estimated service lives obtained for limestone and marble, respectively, in Lisbon. These results are consistent with those of Schouenborg et al. [45], who found that granite is the most durable cladding stone, followed by limestone and then marble, by analysing the strength capacity of stone plates of 200 buildings.

Regarding the colour of the natural stone plates (Figure 10), both light and dark colours exhibit a good correlation between the field results and the degradation curves. Figure 9 shows that the degradation progress of stone claddings with light colours is faster than those with dark colour stones at the preliminary 50 years of the façade's service life in both Tehran and Lisbon. On the contrary, after that time interval, there are no case studies in Tehran, and more data regarding claddings with dark colour stones must be collected in order to obtain unequivocal conclusions. The literature does not provide any indication regarding the relevance of colour for the deterioration of stone claddings [2]. In the Tehran sample, light colour stone claddings tend to present a higher incidence of defects than the dark colour ones, mainly because some defects, such as staining or soot deposition, are easily identified in light colour claddings [14], and, moreover, dark colours are usually associated with more durable stones, such as granites and other eruptive rocks [22].

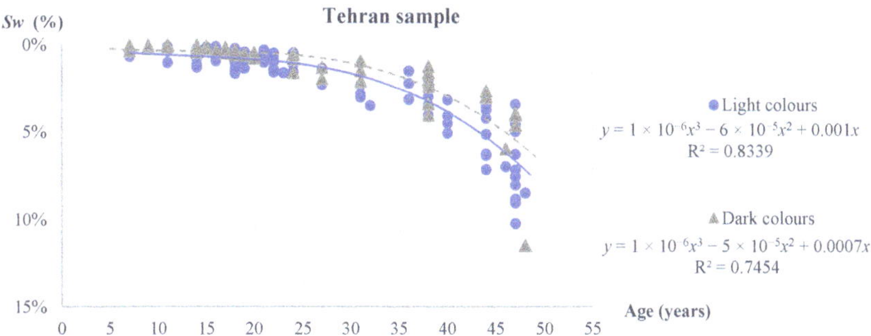

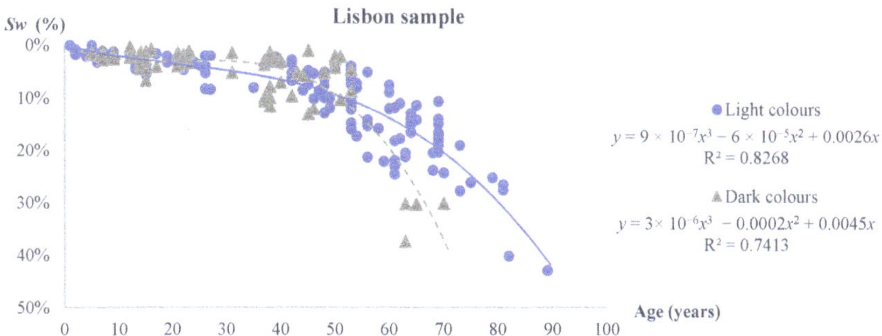

Figure 10. Degradation curves of the Tehran and Lisbon samples, according to the colour of stone.

In terms of type of finishing (Figure 11), subtle variations in the surface finishing produce some differences in decay patterns and the degradation evolution of stone claddings. Various studies suggest that the polishing of stone surfaces would partially inhibit the weathering effect because of the reduction in the open porosity of the stone surface, which exerts an influence on salt decay by fog sea-salt deposition in coastal areas, such as Lisbon, or soot deposition in the polluted air of Tehran [46–48]. In other words, claddings with a rough finishing present higher degradation indexes mainly due to having a higher area of stone exposed to the climatic degradation agents [49]. The results from this study (Figure 11) are coherent with previous findings, and an estimated service life of 67 and 61 years was obtained for claddings with smooth and rough finishings, respectively, in Tehran, while an estimated service life of 69 and 67 years was calculated for Lisbon for the same categories.

Regarding the evolution of the degradation condition of stone claddings according to the size of the stone elements (Figure 12), the results reveal that claddings with medium size plates (area $< 0.4 \text{ m}^2$) present longer estimated service lives than those with larger stone plates (area $\geq 0.4 \text{ m}^2$). As mentioned by Silva et al. [2], this can probably be explained by the fact that the larger the size of the stone elements, the greater the cladding's susceptibility to weather effects, in addition to the lower relative area of the joints in the larger plates and a resulting higher concentration of stresses, which can promote a higher incidence of defects. Based on the degradation curves (Figure 11), an estimated service life of 67 and 61 years was obtained for the Tehran claddings with medium and large stone plates, respectively, and 71 and 66 years for the Lisbon sample.

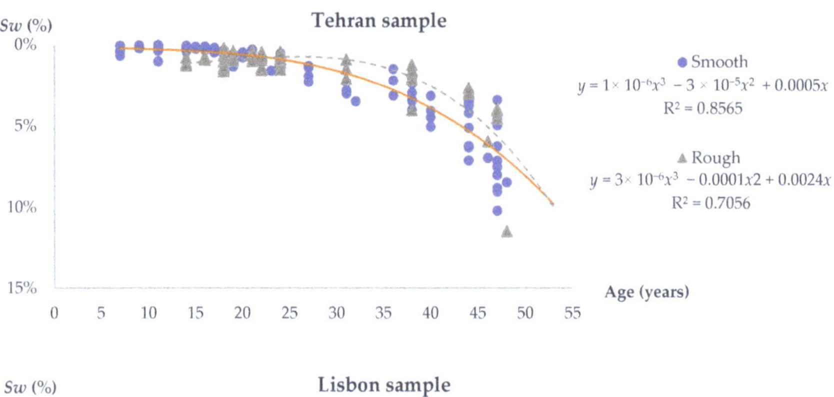

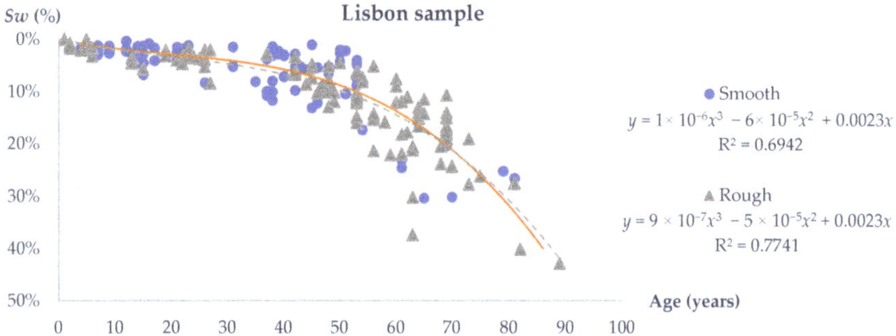

Figure 11. Degradation curves of the Tehran and Lisbon samples, according to the type of finishing.

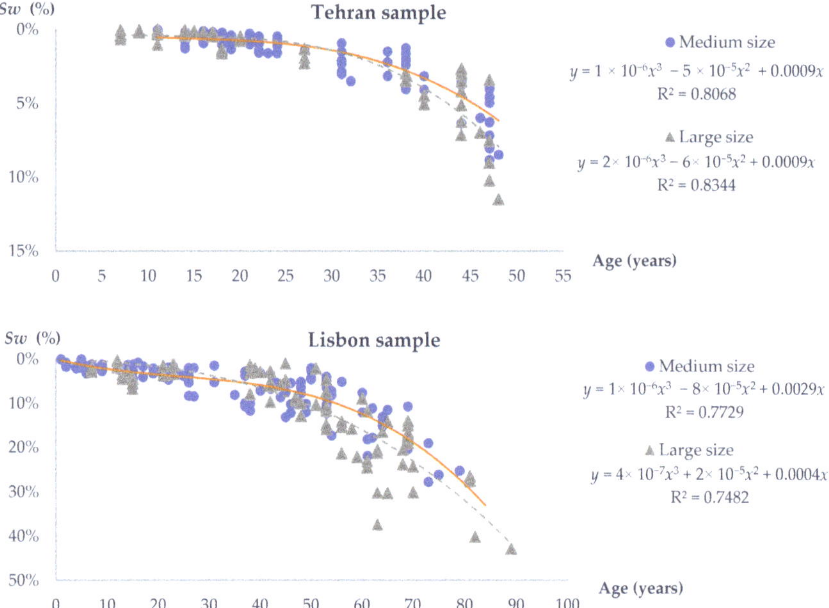

Figure 12. Degradation curves of the Tehran and Lisbon samples, according to the size of the stone elements.

In terms of the location of the cladding (Figure 13), the samples under analysis were grouped depending on whether the stone cladding was located in the taller areas of the façade (integral or partial elevated cladding) or only in the lower floors of the building (partial bottom cladding).

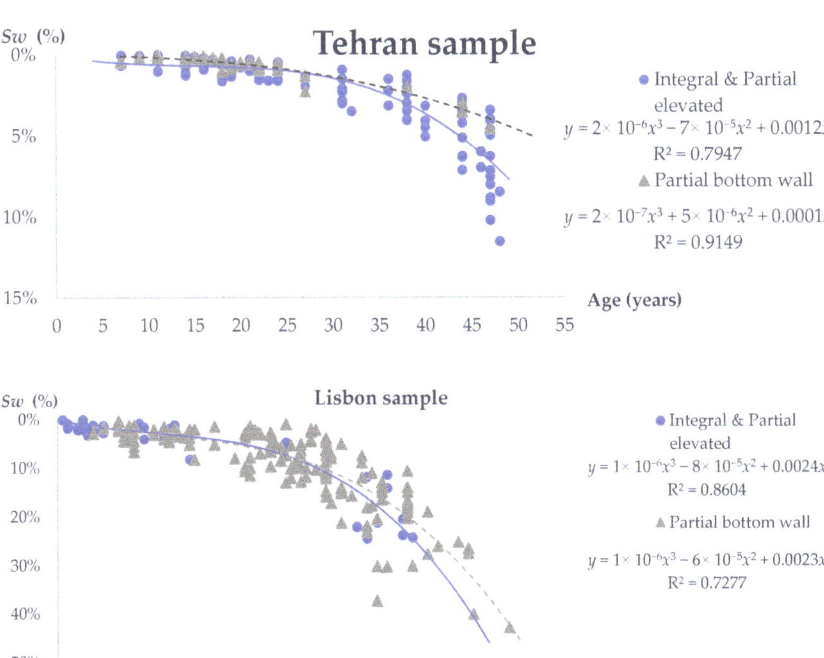

Figure 13. Degradation curves of the Tehran and Lisbon samples, according to the location of the cladding.

Westberg et al. [50] suggested that the higher levels of façades are more exposed to the environmental agents, which can lead to a rapid deterioration of the stone claddings. Furthermore, bottom wall claddings are more accessible to carry out maintenance actions, so they are expected to have longer estimated service lives [2,22]. This study also leads to the same results, as shown in Figure 13, since an estimated service life of 63 years was obtained for partial or elevated claddings and 87 years for bottom wall claddings in the Tehran sample, while an estimated service life of 69 and 65 years was obtained for the bottom wall and elevated stone claddings in Lisbon, respectively.

The literature review on the degradation of building façades reveals that moisture is one of the main factors for materials' deterioration [51]. Wind and solar radiation are responsible factors for moisture settling on the façade and are variable in accordance with the geographical orientation [52]; hence, an analysis of the degradation curves according to the claddings' orientation seems relevant for the analysis of the degradation over time of stone claddings, and they are illustrated in Figure 14.

According to Gaspar and de Brito [14], regarding the rate of the erosion process according to the façades' orientation, the authors concluded that north and west are the most critical orientations in Portugal. This has been confirmed by Emídio et al. [24], who suggested that the stone claddings facing east and south are more durable, while claddings facing west and north tend to present lower service lives. The outcomes of the current study for the Lisbon sample reveal that the most favourable cladding orientations are, in decreasing order, south, east, north and west, with estimated service lives of 71, 70, 67 and 64 years, respectively, which corroborate the findings of the previous works.

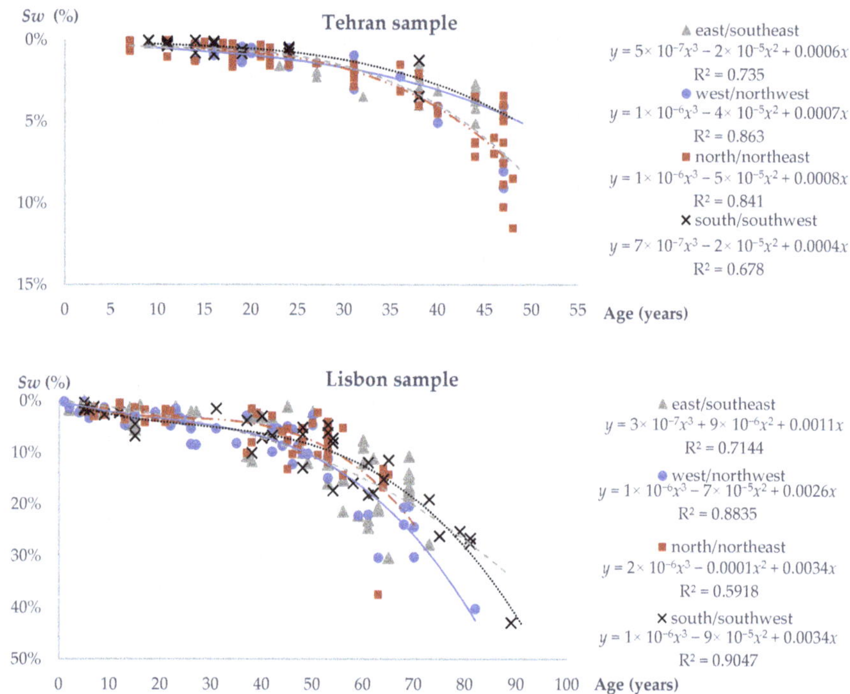

Figure 14. Degradation curves of the Tehran and Lisbon samples, according to the façades' orientation.

In accordance with the meteorological statistics [53], the prevailing winds in Tehran come from the west. Therefore, the probability of the incidence of wind-driven rain is high in that orientation, thus being more prone to suffer from anomalies due to the presence of moisture. Accordingly, in the sample analysed (Figure 14), stone claddings facing north/northeast and west/northwest show lower estimated service lives (63 and 64 years, respectively) in comparison with those facing south/southwest and east/southeast (with estimated service lives of 74 and 78 years, respectively).

Concerning the exposure to the combined action of wind and rain (Figure 15), two categories are considered: (i) stone claddings exposed to a severe action of wind and rain, in façades that are not protected by surrounding buildings or vegetation, in elevated areas; and (ii) stone claddings with a normal or moderate exposure to this action. As mentioned before, in the analysis of the impact of the façades' orientation on the deterioration of stone claddings, the action of driven rain promotes the deterioration of stone claddings, being one of the main causes of erosion and other loss of integrity deficiencies [51,54]. Therefore, stone claddings with a severe exposure to wind–rain action reach the end of their service life sooner (64 and 67 years for Tehran and Lisbon samples, respectively) than claddings with a moderate exposure (85 and 68 years for Tehran and Lisbon samples, respectively).

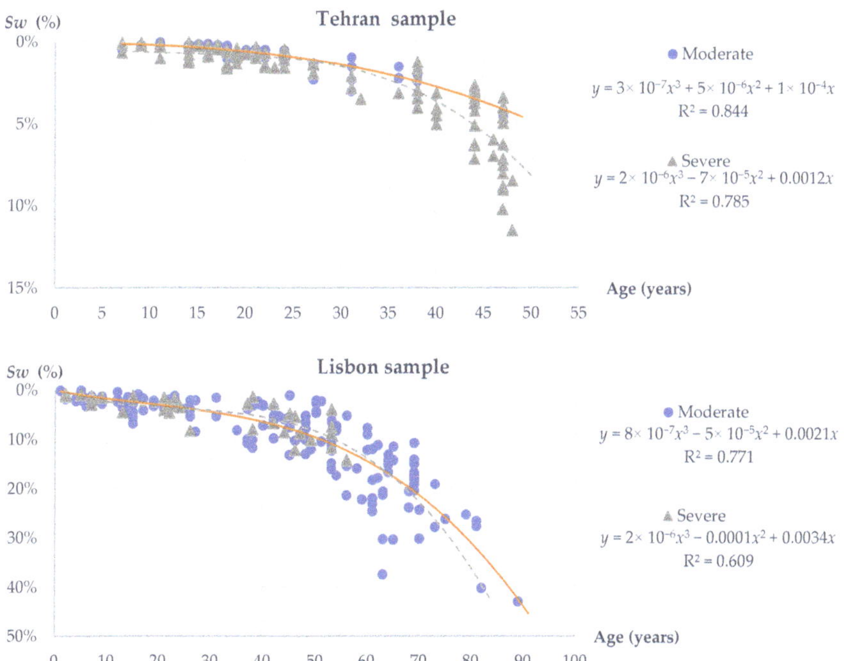

Figure 15. Degradation curves of the Tehran and Lisbon samples, according to the exposure to wind and rain action.

6. Results and Discussion

Table 4 shows a summary of the results obtained for the two samples. In the previous section, as the degradation curves were presented, the results were discussed, and the conclusions of this study can be summarised as follows:

- Tehran, the capital of Iran, is one of the most densely populated cities in the world, with a population around 12 million, and is located in an active seismic zone [55]; for that reason, the structures in Iran are designed for an optimal service life of around 30 years, after which some actions must be carried out to reduce the vulnerability of the buildings to the seismic action [56,57]. In this sense, the Tehran sample presents more recent buildings (with lower ages) than the Lisbon sample (in which the design service life is usually 50 years). According to the proposed model, the natural stone claddings reach the end of their service life after 65 years in Tehran and after 68 years in Lisbon. These values agree with the literature: (i) Silva et al. [2] adopted different statistical models for the service life prediction of stone claddings, and an estimated service life ranging between 68 and 90 years was obtained; (ii) Shohet and Paciuk [41], considering a lower level of users' demand, attained an estimated service life of 64 years (with a range of 59 to 70 years);
- The results obtained for both samples are coherent, and similar degradation patterns were identified in both samples; i.e., stone claddings in more unfavourable conditions reach the end of their service life sooner than claddings more protected from the deterioration agents. The only exception is for the façades' orientation, although for both samples, the north and west orientations have lower service lives due to high exposure to damp and prevailing winds;
- Pitzurra et al. [58] suggest that atmospheric agents and air pollution are the major causes for the degradation of external stone claddings, while other authors [55,59] suggest that the extent of degradation of the stone claddings is influenced by the type of stone used and its mineralogical composition, as well as by its physical and

mechanical properties. The results in Table 4 reveal that the type of stone is the most influential parameter in the service life of natural stone claddings in Lisbon. On the other hand, exposure to prevailing winds and application of the cladding in bottom walls, which are more protected from the environmental agents, seem to be the most relevant factors for the deterioration of stone claddings in Tehran (these characteristics have the highest impact on the loss or gain of the estimated service life of stone claddings in Tehran);

- The results seem to reveal that two different atmospheric agents have a significant impact on the deterioration of stone claddings in the two geographical contexts. In Lisbon, the exposure to damp seems to be a relevant factor, which is accountable by the north orientation; this sample is located in different areas of the city, some closer to the river and the ocean, others more protected, which allows the verification of the impact of this action on the service life of stone claddings. On the other hand, in the Tehran sample, all the claddings are in the same district, thus presenting the same exposure to this action. However, driven rain proved to be a very important degradation agent, mainly because there are also high levels of atmospheric pollution associated with this action. The presence of pollutants was shown to be unquestionably relevant in the Tehran sample, with a higher incidence of anomalies caused by this action, such as the loss of integrity defects (e.g., erosion of stone), discoloration of stone elements and deposition of debris and superficial dirt.

Table 4. Summary of the results obtained for the estimated service life of natural stone claddings in Tehran and in Lisbon.

Cladding Characteristics		R^2 of the Degradation Curves		Estimated Service Life * (Years)		Service Life Gain or Loss According to the Claddings' Characteristics ** (%)	
		Tehran	Lisbon	Tehran	Lisbon	Tehran	Lisbon
Type of stone	Granite	0.6356	0.2008	78	83	20%	22%
	Travertine and limestone	0.8553	0.6865	64	71	−1%	4%
	Marble and crystal	0.6594	0.8350	63	66	−3%	−3%
Colour of natural stone	Light colours	0.8339	0.8268	64	69	−1%	1%
	Dark colours	0.7454	0.7413	67	60	2%	−12%
Type of finishing	Smooth	0.8565	0.6942	66	69	1%	1%
	Rough	0.7056	0.7741	62	67	−5%	−1%
Size of stone plates	Medium size	0.8068	0.7729	67	71	3%	4%
	Large size	0.8344	0.7482	61	66	−5%	−3%
Location of the cladding	Integral	0.7947	0.8604	63	65	−3%	−4%
	Partial bottom	0.9149	0.7277	87	69	34%	1%
Orientation	South/southwest	0.6776	0.9047	74	71	14%	4%
	East/southeast	0.7351	0.7144	78	70	21%	3%
	West/northwest	0.8630	0.8835	64	64	−2%	−6%
	North/northeast	0.8413	0.5918	63	67	−3%	−1%
Wind–rain action	Moderate	0.8442	0.7711	85	68	31%	0%
	Severe	0.7849	0.6090	64	67	−2%	−1%

* The estimated service life is determined using the equations of the degradation curve shown in Figures 7–14. ** These values are obtained through the ratio between the estimated service life for each characteristic and the average estimated service life, which is 65 and 68 years for Tehran and Lisbon samples, respectively.

7. Concluding Remarks

This study analyses the service life and durability of stone claddings in two environmental contexts, namely in Lisbon, Portugal, and in Tehran, Iran. The methodology applied in this study was initially developed for rendered façades in Lisbon [14], with the aim of being a general framework for predicting the service life of building elements. In this sense, this empirical method was adapted and applied to other claddings, in particular to natural stone claddings [2,22]. In the previous approaches, the authors mentioned that this method could be applied to other elements and to other geographical contexts. In fact, even though there are significant differences between the two locations, the proposed methodology proves to be able to accurately predict the service life of stone claddings in Tehran according to their intrinsic properties, such as the type of stone, their design conditions and their exposure to environmental agents.

Similar degradation patterns were identified in both samples in accordance with the physical deterioration process of natural stone claddings over time. The results obtained revealed that the characteristics of stone claddings strongly influence their service life. The climatic or atmospheric agents present a significant impact on the deterioration of stone claddings. However, the most conditioning climatic agents for the deterioration of stone claddings seem to change in the two geographical contexts. In Lisbon, the exposure to damp seems to be the most conditioning factor, while in Tehran, the exposure to driven rain appears to be a very significant degradation agent, mainly because there are also high levels of atmospheric pollution associated with this action.

This study proposes an empirical tool to evaluate the service life of natural stone claddings, considering their on-site performance. The application of this method to two different locations proves the validity and applicability of the tool to different contexts. This study provides some knowledge about the degradation and service life of stone claddings, quantifying the variability of the estimated service life introduced by the different characteristics analysed. This information can be extremely useful both in design and maintenance stages for the adoption of more rational and durable solutions.

Author Contributions: Conceptualisation, S.H.M., A.S. and J.d.B.; methodology, S.H.M. and A.S.; formal analysis, S.H.M., A.S. and J.d.B.; investigation, S.H.M. and A.S.; data curation, S.H.M. and A.S.; writing—original draft preparation, S.H.M. and A.S.; writing—review and editing, J.d.B., A.E. and S.B.H. All authors have read and agreed to the published version of the manuscript.

Funding: This research was funded by the Portuguese Foundation for Science and Technology (FCT) through project BestMaintenance-LowerRisks (PTDC/ECI-CON/29286/2017).

Institutional Review Board Statement: Not applicable.

Informed Consent Statement: Not applicable.

Data Availability Statement: The data presented in this study are available on request from the corresponding author.

Acknowledgments: The authors gratefully acknowledge the support of the CERIS Research Centre (Instituto Superior Técnico-University of Lisbon).

Conflicts of Interest: The authors declare no conflict of interest.

References

1. Wekesa, B.; Steyn, G.; Otieno, F. The response of common building construction technologies to the urban poor and their environment. *Build. Environ.* **2010**, *45*, 2327–2335. [CrossRef]
2. Silva, A.; De Brito, J.; Gaspar, P.L. Service life and durability of assemblies. In *Methodologies for Service Life Prediction of Buildings*; Springer: Berlin/Heidelberg, Germany, 2016; pp. 13–66.
3. Haagenrud, S. Factors causing degradation. In *Guide and Bibliography to Service Life and Durability Research for Buildings and Components*; CIB Working Commission W080/RILEM Technical Committee 140-TSL: Ottawa, ON, Canada, 2004; p. 108.
4. Chai, C.; De Brito, J.; Gaspar, P.; Silva, A. Predicting the service life of exterior wall painting: Techno-economic analysis of alternative maintenance strategies. *J. Constr. Eng. Manag.* **2013**, *140*, 04013057. [CrossRef]
5. Daniotti, B.; Spagnolo, S. Service life prediction tools for buildings' design and management. In Proceedings of the 11th International Conference on Durability of Building Materials and Components (11th DBMC), Istanbul, Turkey, 11–14 May 2008.
6. Pearce, D.W. The social and economic value of construction. In *The Construction Industry's Contribution to Sustainable Development, NCRISP*; Davis Langdon Consultancy: London, UK, 2003.
7. Brand, S. *How Buildings Learn: What Happens after They're Built?* 1st ed.; Phoenix Illustrated: London, UK, 1997.
8. Slaughter, E.S. Design strategies to increase building flexibility. *Build. Res. Inf.* **2001**, *29*, 208–217. [CrossRef]
9. Gaspar, P.L. Service Life of Constructions: Development of a Methodology to Estimate the Durability of Construction Elements, Application to Renders in Current Buildings. Ph.D. Thesis, Instituto Superior Técnico, Technical University of Lisbon, Lisbon, Portugal, 2009. (In Portuguese)
10. Neto, N.; De Brito, J. Inspection and defect diagnosis system for natural stone cladding. *J. Mater. Civ. Eng.* **2011**, *23*, 1433–1443. [CrossRef]

11. Watt, D.S. *Building Pathology: Principles and Practice*, 2nd ed.; Blackwell Science Ltd., Blackwell Publishing Company: London, UK, 2007.

12. Rivard, H.; Bédard, C.; Fazio, P.; Ha, K.H. Functional analysis of the preliminary building envelope design process. *Build. Environ.* **1995**, *30*, 391–401. [CrossRef]

13. Yılmaz, N.G.; Goktan, R.M.; Kibici, Y. Relations between some quantitative petrographic characteristics and mechanical strength properties of granitic building stones. *Int. J. Rock Mech. Min. Sci.* **2011**, *48*, 506–513. [CrossRef]

14. Gaspar, P.L.; De Brito, J. Quantifying environmental effects on cement-rendered facades: A comparison between different degradation indicators. *Build. Environ.* **2008**, *43*, 1818–1828. [CrossRef]

15. Norvaišienė, R.; Miniotaitė, R.; Stankevičius, V. Climatic and air pollution effects on building facades. *Mater. Sci.* **2003**, *9*, 102–105.

16. Frohnsdorff, G.; Martin, J. Towards prediction of building service life: The standards imperative. In Proceedings of the 7th Durability of Building Materials and Components (7th DBMC), Stockholm, Sweden, 19–23 May 1996.

17. Kus, H.; Carlsson, T. Microstructural investigations of naturally and artificially weathered autoclaved aerated concrete. *Cem. Concr. Res.* **2003**, *33*, 1423–1432. [CrossRef]

18. Searls, C.L.; Thomasen, S.E. Repair of the terra-cotta facade of Atlanta City Hall. In Proceedings of the 2nd International Conference on Structural Repair and Maintenance of Historical Buildings, Seville, Spain, 14–16 May 1991.

19. Henriksen, J.F. Reactions of gases on calcareous stones under dry conditions in field and laboratory studies. *Water Air Soil Pollut.* **1995**, *85*, 2707–2712. [CrossRef]

20. Shohet, I.; Rosenfeld, Y.; Puterman, M.; Gilboa, E. Deterioration patterns for maintenance management—A methodological approach. In Proceedings of the 8th International Conference on Durability of Building Materials and Components, Vancouver, BC, Canada, 30 May–3 June 1999.

21. Cohen, L.; Manion, L.; Morrison, K. Ex post facto research. In *Research Methods in Education*, 6th ed.; Taylor & Francis Group: Oxfordshire, UK, 2007; pp. 267–271.

22. Silva, A.; De Brito, J.; Gaspar, P.L. Service life prediction model applied to natural stone wall claddings (directly adhered to the substrate). *Constr. Build. Mater.* **2011**, *25*, 3674–3684. [CrossRef]

23. Silva, A.; Dias, J.L.; Gaspar, P.L.; De Brito, J. Statistical models applied to service life prediction of rendered façades. *Autom. Constr.* **2013**, *30*, 151–160. [CrossRef]

24. Emídio, F.; De Brito, J.; Gaspar, P.L.; Silva, A. Application of the factor method to the estimation of the service life of natural stone cladding. *Constr. Build. Mater.* **2014**, *66*, 484–493. [CrossRef]

25. Ximenes, S.; de Brito, J.; Gaspar, P.L.; Silva, A. Modelling the degradation and service life of ETICS in external walls. *Mater. Struct.* **2015**, *48*, 2235–2249. [CrossRef]

26. Silva, A.; Gaspar, P.L.; De Brito, J.; Neves, L.C. Probabilistic analysis of degradation of façade claddings using Markov chain models. *Mater. Struct.* **2016**, *49*, 2871–2892. [CrossRef]

27. Mousavi, S.H.; Silva, A.; De Brito, J.; Ekhlassi, A.; Hosseini, S.B. Service life prediction of natural stone claddings with an indirect fastening system. *J. Perform. Constr. Facil.* **2017**, *31*, 04017014. [CrossRef]

28. Mousavi, S.H.; Silva, A.; De Brito, J.; Ekhlassi, A.; Hosseini, S.B. Influence of design on the service life of indirectly fastened natural stone cladding. *J. Perform. Constr. Facil.* **2019**, *33*, 04019021. [CrossRef]

29. Mousavi, S.H. The Appraisal Model of Architectural Design of Natural Stone Cladding in Order to Predict its Service Life. Ph.D. Thesis, Iran University of Science & Technology, Tehran, Iran, 2018. (In Persian)

30. Camposinhos, R.S. Natural stone characterization. In *Stone Cladding Engineering*; Springer: Dordrecht, The Netherlands, 2014; pp. 9–33.

31. Camposinhos, R.S. Wall and cladding systems. In *Stone Cladding Engineering*; Springer: Dordrecht, The Netherlands, 2014; pp. 37–54.

32. Köppen-Geiger. World Map of the Köppen-Geiger Climate Classification. 2017. Available online: http://koeppen-geiger.vu-wien.ac.at/present.htm (accessed on 16 November 2020).

33. Shohet, I.M.; Laufer, A. Exterior cladding methods: A technoeconomic analysis. *J. Constr. Eng. Manag.* **1996**, *122*, 242–247. [CrossRef]

34. Bauer, E.; De Freitas, V.P.; Mustelier, N.; Barreira, E.; De Freitas, S.S. Infrared thermography—Evaluation of the results reproducibility. *Struct. Surv.* **2015**, *33*, 20–35. [CrossRef]

35. Meola, C.; Di Maio, R.; Roberti, N.; Carlomagno, G.M. Application of infrared thermography and geophysical methods for defect detection in architectural structures. *Eng. Fail. Anal.* **2005**, *12*, 875–892. [CrossRef]

36. Atash, F. The deterioration of urban environments in developing countries: Mitigating the air pollution crisis in Tehran, Iran. *Cities* **2007**, *24*, 399–409. [CrossRef]

37. Bayat, R.; Ashrafi, K.; Motlagh, M.S.; Hassanvand, M.S.; Daroudi, R.; Fink, G.; Künzli, N. Health impact and related cost of ambient air pollution in Tehran. *Environ. Res.* **2019**, *176*, 108547. [CrossRef]

38. Lamhasni, T.; El-Marjaoui, H.; El Bakkali, A.; Lyazidi, S.A.; Haddad, M.; Ben-Ncer, A.; Benyaich, F.; Bonazza, A.; Tahri, M. Air pollution impact on architectural heritage of Morocco: Combination of synchronous fluorescence and ATR-FTIR spectroscopies for the analyses of black crusts deposits. *Chemosphere* **2019**, *225*, 517–523. [CrossRef]

39. Ortega-Morales, O.; Montero-Muñoz, J.L.; Neto, J.A.B.; Beech, I.B.; Sunner, J.; Gaylarde, C. Deterioration and microbial colonization of cultural heritage stone buildings in polluted and unpolluted tropical and subtropical climates: A meta-analysis. *Int. Biodeterior. Biodegrad.* **2019**, *143*, 104734. [CrossRef]

40. Vieira, S.M.; Silva, A.; Sousa, J.M.C.; De Brito, J.; Gaspar, P.L. Modelling the service life of rendered facades using fuzzy systems. *Autom. Constr.* **2015**, *51*, 1–7. [CrossRef]

41. Shohet, I.M.; Paciuk, M. Service life prediction of exterior cladding components under standard conditions. *Constr. Manag. Econ.* **2004**, *22*, 1081–1090. [CrossRef]

42. Lounis, Z.; Vanier, D.; Lacasse, M.; Kyle, B. Decision-support system for service life asset management: The BELCAM project. In Proceedings of the 8th International Conference on Durability of Building Materials and Components, Vancouver, BC, Canada, 30 May–3 June 1999.

43. Prieto, A.; Silva, A. Service life prediction and environmental exposure conditions of timber claddings in South Chile. *Build. Res. Inf.* **2019**, *48*, 191–206. [CrossRef]

44. Souza, J.; Silva, A.; De Brito, J.; Bauer, E. Analysis of the influencing factors of external wall ceramic claddings' service life using regression techniques. *Eng. Fail. Anal.* **2018**, *83*, 141–155. [CrossRef]

45. Schouenborg, B.; Grelk, B.; Malaga, K. Testing and assessment of marble and limestone (TEAM)—Important results from a large European research project on cladding panels. *J. ASTM Int.* **2007**, *4*, 1–14. [CrossRef]

46. Chew, M.; Tan, P. Facade staining arising from design features. *Constr. Build. Mater.* **2003**, *17*, 181–187. [CrossRef]

47. Benavente, D.; Garcı, M.; Garcí, J.; Sánchez-Moral, S.; Ordóñez, S. Role of pore structure in salt crystallisation in unsaturated porous stone. *J. Cryst. Growth* **2004**, *260*, 532–544. [CrossRef]

48. Urosevic, M.; Sebastián, E.; Cardell, C. An experimental study on the influence of surface finishing on the weathering of a building low-porous limestone in coastal environments. *Eng. Geol.* **2013**, *154*, 131–141. [CrossRef]

49. Siedel, H.; Siegesmund, S.; Sterflinger, K. Characterisation of stone deterioration on buildings. In *Stone in Architecture*; Springer: Berlin/Heidelberg, Germany, 2011; pp. 347–410.

50. Westberg, K.; Norén, J.; Kus, H. On using available environmental data in service life estimations. *Build. Res. Inf.* **2001**, *29*, 428–439. [CrossRef]

51. Pereira, C.; De Brito, J.; Silvestre, J.D. Contribution of humidity to the degradation of façade claddings in current buildings. *Eng. Fail. Anal.* **2018**, *90*, 103–115. [CrossRef]

52. Nascimento, M.; Bauer, E.; De Souza, J.; Zanoni, V. Wind-driven rain incidence parameters obtained by hygrothermal simulation. *J. Build. Pathol. Rehabil.* **2016**, *1*, 5. [CrossRef]

53. Windfinder. Wind & Weather Statistics: Tehran. 2017. Available online: https://www.windfinder.com/windstatistics/tehran_mehrabad-airport (accessed on 16 November 2020).

54. Camuffo, D. Physical weathering of stones. *Sci. Total. Environ.* **1995**, *167*, 1–14. [CrossRef]

55. Panahi, M.; Rezaie, F.; Meshkani, S.A. Seismic vulnerability assessment of school buildings in Tehran city based on AHP and GIS. *Nat. Hazards Earth Syst. Sci.* **2014**, *14*, 969–979. [CrossRef]

56. Zekai, S. Supervised fuzzy logic modelling for building earthquake hazard assessment. *Expert Syst. Appl.* **2011**, *38*, 14564–14573.

57. Sharifzadegan, M.H.; Fathi, H. Application of seismic risk assessment models in urban planning and design. *Soffeh* **2008**, *17*, 109–124.

58. Pitzurra, L.; Moroni, B.; Nocentini, A.; Sbaraglia, G.; Poli, G.; Bistoni, F. Microbial growth and air pollution in carbonate rock weathering. *Int. Biodeterior. Biodegrad.* **2003**, *52*, 63–68. [CrossRef]

59. Prikryl, R.; Lokajicek, T.; Svobodova, J.; Weishauptova, Z. Experimental weathering of marlstone from Přední Kopanina (Czech Republic)—Historical building stone of Prague. *Build. Environ.* **2003**, *38*, 11. [CrossRef]

 *buildings*

Article

Most Frequent Problems of Building Structures of Urban Apartment Buildings from 2nd Half of 19th Century and the Start of 20th Century

Klara Kroftova

Faculty of Civil Engineering, Czech Technical University in Prague, 166 36 Prague 6, Czech Republic; Klara.Kroftova@fsv.cvut.cz

Abstract: An urban residential building from the second half of the 19th century and the start of the 20th century, the so-called tenement house, is a significant representative of the architecture of the developing urban fabric in Central Europe. The vertical and horizontal load-bearing structures of these houses currently tend to show characteristic, repeated defects and failures. Their knowledge may, in many cases, facilitate and speed up the design of the historic building's restoration without compromising its heritage value in this process. The article presents the summary of the most frequently occurring defects and failures of these buildings. The summary, however, is not an absolute one, and, in the case of major damage to the building, it still applies that, first of all, a detailed analysis of the causes and consequences of defects and failures must be made as a basic prerequisite for the reliability and long-term durability of the building's restoration and rehabilitation. An integral part of the rehabilitation of buildings must be the elimination of the causes of the appearance of their failures and remediation of all defects impairing their structural safety, health safety and energy efficiency.

Keywords: apartment building; 19th century; horizontal structure; vertical structure; defects; failures

Citation: Kroftova, K. Most Frequent Problems of Building Structures of Urban Apartment Buildings from 2nd Half of 19th Century and the Start of 20th Century. *Buildings* **2021**, *11*, 27. https://doi.org/10.3390/buildings 11010027

Received: 8 December 2020
Accepted: 1 January 2021
Published: 12 January 2021

Publisher's Note: MDPI stays neutral with regard to jurisdictional claims in published maps and institutional affiliations.

1. Introduction

The apartment building became a new phenomenon in Central Europe in the 19th century, although rental flats had been quite common in historic cities, practically from the Middle Ages. However, the situation in individual cities differed. Architect Pavel Janak summed up the situation in his article "One Hundred Years of a Residential Rental House in Prague" (1933): "The situation in Prague and our towns in the 19th century in general is characterized by the non-existence of a family house. Towns are made up of large houses with many flats, built by individuals and rented out. Our culture, or its lack, is based on collective housing in a large building. A family house and family living is not a component of our culture. In the second half of the 19th century, the type of a large house became the subject of unprecedented business. The primary reason why it was built was not to provide flats, but to make money fast. The size of the house and its massive scale were only paralyzed at the end of the 19th century by increased demands for the value of an apartment—leading to the shrinkage of the house to only two apartments on one floor and to the appearance of a family house, a hitherto unknown type of housing" [1].

The courtyard tracts of historic houses represent an initial early stage in the development of apartment buildings in Czech cities—while the solidly built foundation of a house with a vaulted ground floor was often left even during a radical reconstruction, the newly built courtyard tracts connected with the house by galleries emerged as completely new structures [2]. In Prague (the capital city of the Czech Kingdom of the Austro-Hungarian Empire of the time) three or, less often, even four-storey houses were built in the busiest streets on the threshold of the 19th century (Figure 1). On side streets, the houses mostly had two to three storeys, and single-storey houses were an exception. The houses had courtyard tracts and the space was often built-up to the maximum extent. The situation differed in other large Czech cities (e.g., Pilsen or Ostrava), where three-storey houses were a rarity; moreover, they were only built in the main squares. Two-storey houses, and single-storey houses in side streets, were much more common there. The former historic royal towns, which usually had around 220 to 250 houses in the "walls", used to have mostly single-storey houses, and a two-storey house was an exception. At the same time, the smaller the town, the fewer the number of rental flats.

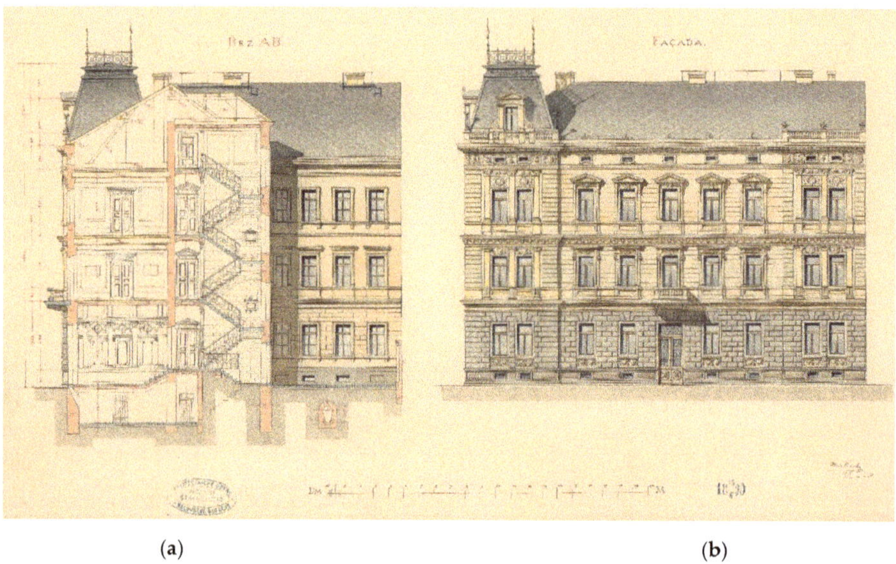

(**a**)(**b**)

Figure 1. Section (**a**) and elevation (**b**) of a typical four-storey tenement house from the end of the 19th century on the student work of A. Censky, 1890 [3].

Until the beginning of the 18th century, there were no binding rules in force for the design of buildings. The fundamental construction principles were based on the builders' efforts to create the required interior space volume defined by vertically laid-out load-bearing structures, divided horizontally by floor structures and enclosed by the roof.

From about the mid-18th to the mid-19th century, the most important areas related to building law were refined, mainly in connection to fire safety. The buildings were designed on the basis of builders' experience where the thickness of vertical load-bearing structures was determined by empirical formulae depending on the number of storeys, the height of the storeys and the depth of the tract. During the 19th century, three waves of building codes were issued, which partly reflected the development of urban construction practice. The first wave includes the Czech Building Regulations of 1833, the Moravian and Silesian Building Regulations of 1835 and the amended version of the Czech Building Regulations of 1845. The second wave may be represented by the Czech Building Regulations of 1864, the Moravian Building Regulations of 1869 and the Silesian Building Regulations of 1883. The third wave covers the Prague Building Regulations of 1886, which were adopted by Pilsen and České Budějovice in 1887, the Czech Building Regulations of 1889 applying to the rest of the country and two Moravian Building Regulations of 1894 [4].

2. Materials and Methods

Vertical and horizontal load-bearing structures of apartment buildings from the second half of the 19th century currently show, in many cases, characteristic, recurring defects and faults (Figure 2). The main goal of this paper is to identify the most common defects and failures of masonry structures based on field and archival research of building structures and details of masonry buildings from the period. Knowledge of endangered structural details of vertical and horizontal structures, including, for example, the issue of basic construction conditions, is a suitable step to mitigate the consequences of construction and technical problems of historic buildings. At the same time, their knowledge can facilitate and speed up the design of the restoration of a historic building so that no monumental values are lost during this process [5].

(a)

(b)

Figure 2. Images clearly showing the most common failures of structures from the second half of the 19th and the early 20th centuries caused by degradation processes initiated in these cases by the combination of moisture, an incorrect design solution and neglected maintenance. The following are clearly visible on images: (**a**) fallen plaster, damaged ledge by massive leakage, mold and mosses; (**b**) degradation and disruption of plaster, significant surface efflorescence of salts, mold and mosses.

A summary of the most common defects and failures of these buildings, presented in the following parts of the article, is based on field research, especially of Prague apartment buildings. The archival research is based on historical building literature of the 19th century and a search of drawing documentation of students of construction schools, which is available at the National Technical Museum in Prague [6].

2.1. Vertical Structures

The primary structural function of vertical masonry structures is to transfer the effects of vertical and horizontal loads from individual floor structures through the foundation structure to the foundation subsoil. Tenement houses from the 19th century, but also shops, schools, hospitals, public buildings, etc., were characterized by the longitudinal layout of load-bearing walls, parallel to the street tract. Inside the building, the load-bearing walls formed the so-called middle walls, where the depth of the tracts most often ranged between 4 and 5 m, exceptionally up to 8 m (palaces, public buildings), while internal tracts were usually less than 4 m deep (Figure 3a). The vertical load-bearing structures of urban apartment buildings were most often executed in mixed masonry and, above all, masonry of solid burnt bricks of the so-called classic format of 290 mm × 140 mm × 65 mm [7,8]. In exceptional cases, the walls of overground storeys in buildings from this period could already be built of hollow burnt-clay bricks or hollow brick-like blocks of lightweight concrete (slag concrete or slag pumice), or blocks of another "modern" material. The end of the 19th century was marked by an increase in the use of rolled iron elements (beams, columns, etc.), to be later followed by a global development of concrete and reinforced concrete structures.

Pursuant to the above building codes, urban residential masonry buildings were characterized by massive walls 300–900 mm thick, where, starting from the second wave of building regulations (i.e., from the second half of the 19th century), a minimum wall thickness of 450 mm was required to prevent frost penetration and was further increased in a downward direction (Figure 3b). Lime or cement-lime mortar was used as the most common bonding material, while the use of cement mortar increased at the end of the century, and, subsequently, in the 20th century.

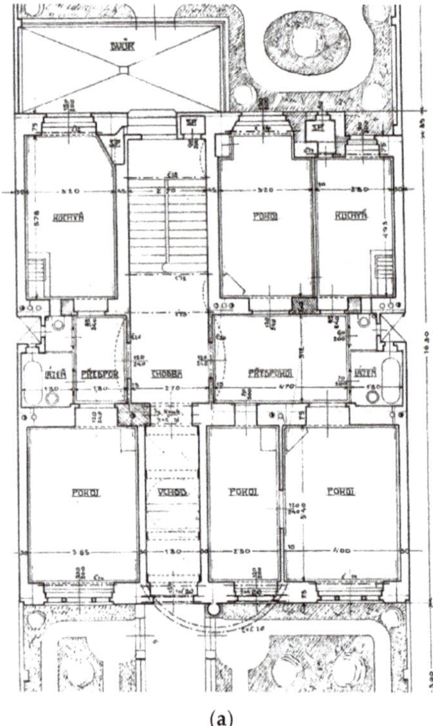
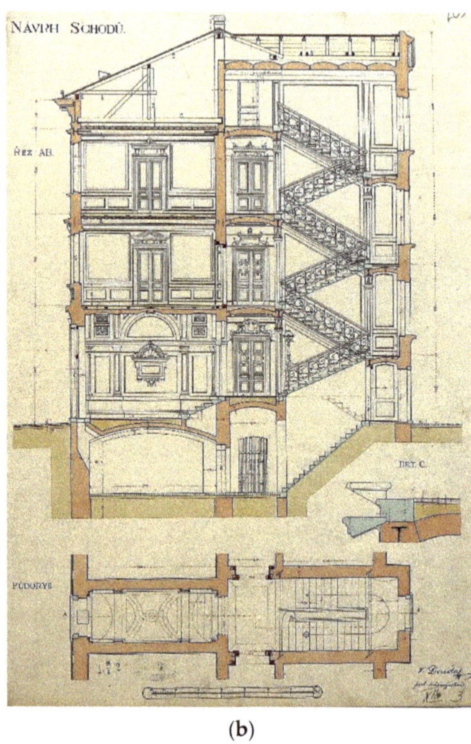

Figure 3. The floor plan and section of urban apartment buildings: (**a**) floor plan of house with four and two-room apartment, both are equipped with bathrooms (arch. V. Reznicek, early 20th century) [9]; (**b**) part of a four-storey house with a clearly legible arrangement of the vertical supporting structure with increasing thickness downwards (V. Douda, 1902) [10].

The stability of the load-bearing system of masonry buildings with several storeys was ensured by longitudinal—front and middle—walls, on which floor structures and transverse walls properly bonded to the longitudinal walls were mounted. The transversely laid-out walls—gable walls, staircase walls, brick partitions 150 mm and 300 mm in thickness—coupled the longitudinal, primarily load-bearing walls, ensuring their stability and contributing to the spatial performance of the load-bearing system, and played an important role in enhancing stability.

The so-called wall and beam ties fulfil an important function in terms of ensuring spatial stiffness, stability and resistance to the effects of forced deformations (Figure 4). From the beginning of the 19th century, wrought iron wall ties were used in walls and vaults, where they served to stiffen and clamp the building and absorb oblique compressive loads, particularly during masonry settling and the additional compression of foundation soil.

2.2. Horizontal Structures

In the majority of urban apartment buildings from the second half of the 19th and the start of the 20th century, various types of floor structures can be found. From a fireproof point of view, ceiling structures are one of the most important parts of the building. At the same time, the fire safety properties of horizontal structures not only reflect their historical development, but also represent an important criterion in categorization. "Numerous construction ceilings can be divided into more or less completely combustion—if it had used exclusively wood because besides still other substances—and incombustible in which wood is unfit on at all." [7] (p. 72).

With regard to the ability of horizontal structures to resist the effects of fire, ceilings of the monitored period can be divided into:

- **combustible floor structures** (e.g., a simple wooden beamed ceiling with a flap, i.e., with visible wooden beams),
- **semi-combustible floor structures**, fitted with a ceiling embankment of at least 80 mm thick at the upper face, while at the lower face, the wooden beam structure is usually protected by undercover of planks covered by reed plaster (e.g., beam ceiling with reed plaster, flap and ceiling embankment etc.),
- **non-combustible floor structures**, i.e., made of bricks, concrete, cast or wrought iron or later mild steel, " . . . in which wood is not used at all. . . . A safe, non-combustible ceiling is provided by a vault. Large spaces cannot be bridged by individual arches for several reasons, mainly because they require strong retaining walls, narrow the spaces and are expensive. The effort to eliminate these defects leads to division of by structures serving as support for smaller vaults. First, it tracks the vault, then the iron beams of either rolling surface (traverse) or riveted (tin) in shape, providing at least a quantity of the spent materials greatest resistance against bending." [7] (p. 77).

According to current knowledge, the above-mentioned division of ceiling structures, from the point of view of fire resistance, is overly simplified and inaccurate [11,12]. Nevertheless, at the time of formulating the above-cited building regulations, it was a significant step to increase the safety of buildings, i.e., also city apartment buildings.

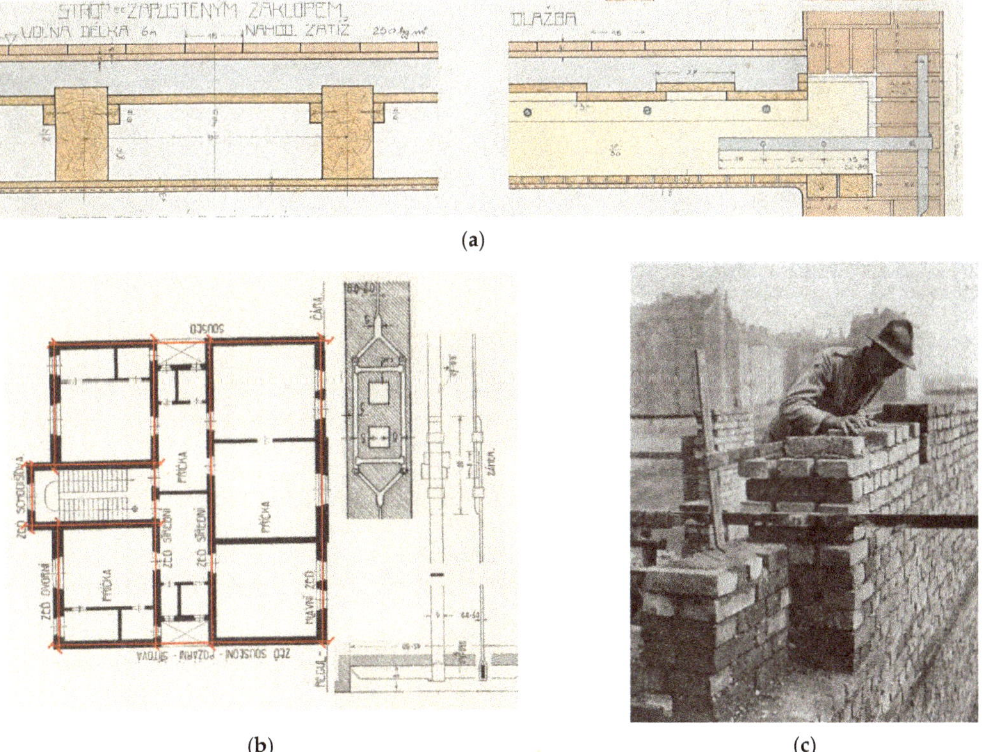

Figure 4. The wall and beam ties: (**a**) the beam wrought iron ties in the drawing on the school work of V. Vašatko from 1910 [13]; (**b**) the detail of the wall ties and their location (red lines) in the floor plan of the house [14]; (**c**) wall ties made of wrought iron, including adjustment with a lock, on a period photograph (1932) when performing a masonry structure [15].

The dominating types of horizontal structures used were combustible and semi-combustible floor structures. Specific types of floors include, in particular, wooden beam floors with subflooring and, later, also with reed plaster, or an upright plank floor with diagonal braces with reed plaster, subflooring and an embankment, etc.

In rooms with strict fire safety requirements, i.e., mainly in public buildings, but also in some areas of apartment buildings (e.g., the ground floor), non-combustible floor structures were mostly designed. In the apartment buildings of the 19th century, vaults were considered to be a safe, non-combustible ceiling structure (or were required by law and building regulations for the Czech Kingdom). The vaults were used (or required by the building codes) in gateways, entrances to houses, cellars and potentially, if necessary, in staircases, corridors and on the ground floor of apartment buildings were considered as safe, non-combustible floor structures. They were usually of a wagon vault type with arch rings, and, later, flat mirrored vaults were used. In this period, the non-combustible ceiling structures used also included segmental vaults made of solid fired bricks vaulted between the beams made from cast iron, wrought iron or later steel (Figure 5). The structures used at the turn of the 19th and early 20th centuries also included various types of masonry floor structures derived from a barrel vault, so-called straight or flat vaults made up of ceramic masonry elements, and, later, of plain or reinforced concrete.

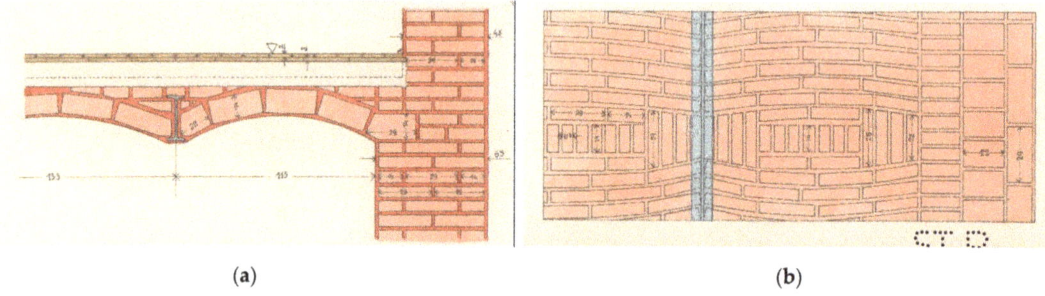

(a)   (b)

Figure 5. Detail of a vaulted ceiling in a school project, unknown author, 1920s. Section (**a**) and top view (**b**) of a brick vault into traverses with a clearly visible composition of bricks [16].

3. Results

The main causes of degradation processes, which generally lead to the deterioration of the required properties of materials and structures, are, on the one hand, specific properties of building materials (their composition, structure, etc.) and, on the other hand, time-variable parameters of the exterior environment in which buildings and their parts perform [17,18]. Different degradation processes caused by climatic factors differ from each other in their duration and type of effect. Some last for fractions of a second, while others for days, months or even years. These parameters, together with material parameters, create the conditions that will initiate or accelerate mechanical, mineralogical, physical, chemical and biological degradation processes [19]. It may be summarized that the defects and failures of structures from the second half of the 19th and the start of the 20th century are most frequently caused by design defects, material defects, by degradation processes triggered mainly by moisture, by incorrect design solutions, faulty workmanship, usage and by neglected maintenance [20–22].

3.1. Vertical Structures

A frequent cause of failures of vertical masonry structures of buildings from the second half of the 19th and the start of the 20th century, which are manifested by cracks, crushing and spalling of surface layers, etc., is poor quality and strength of the masonry in compression and tension plus some other effects, such as different properties of the masonry elements made of mixed masonry or multi-leaf masonry, respectively, low compressive and tensile strength of masonry elements and mortar, etc. [23]. The quality of the masonry and its load-bearing capacity result from the quality of the masonry elements, the bonding mortar and masonry workmanship, but, in particular, from compliance with the principles of proper masonry bonding [24]. The quality of masonry is also significantly affected by increased moisture levels or its aggressiveness and associated degradation processes.

Defects in masonry are caused by imperfect project documentation or imperfect or incorrect execution. The most common execution defect is non-compliance with the principles of proper bonding. The defects in masonry subsequently manifested by the occurrence of failures, insufficient load-bearing capacity, low resistance to degradation processes and others, include:

- poor quality of mortar (composition, excessive grain size of aggregates, low adhesion, shrinkage, poor workability, non-homogeneity) [25],
- poor quality of masonry elements (excessive porosity, insufficient compressive and tensile strength, high moisture absorption, differences in the quality of individual pieces of building material—masonry elements) [26,27],
- non-observance of the planeness and verticality of masonry elements.

In the case of buildings from the second half of the 19th and the start of the 20th century, the cause of these failures resulting from a change in the footing bottom shape is an error in their project preparation, which lacked a geological survey of foundation conditions. At the same time, the empirical design of building structures and insufficient knowledge of the above-mentioned issues led to the construction of foundation structures with insufficient stiffness, without expansion joints and to insufficient reinforcement of the load-bearing masonry at the level of floor structures (by wall and beam ties, bond beams, tie rods, etc.).

The failures of vertical load-bearing masonry structures are crucial in terms of ensuring the mechanical resistance and stability of the load-bearing system. Structural (mechanical) failures of masonry structures and buildings are caused by the response of the building structure to static loads and the deformation effect of some loads and non-force effects, or to static loads with a dynamic component. The most common manifestations of these failures are excessive deformations and strains, cracks (failure in tension and shear), crushing (failure in compression) and local mechanical damage arising due to insufficient load-bearing capacity, in places of stress concentrations, as a result of a change in loading, long-term deformations, degradation and disintegration of masonry, or due to unprofessional interventions. (Figure 6) [28–30]. These failures are directly related to the structural behaviour of the load-bearing structure and affected by their occurrence in the structural performance of the load-bearing system, either locally or as a whole—they are critical in terms of ensuring the mechanical resistance and stability of the supporting system [31,32].

Non-structural failures of masonry structures and buildings are caused by the interaction of materials used for individual structures with the external environment, manifested, above all, by increased moisture levels, chemical, mineralogical and biological processes which weaken the required properties of structures and cause their gradual deterioration and disintegration—degradation processes [33].

The failures of masonry are closely related to the failures of the adjoining masonry structures—vaults, springers and lintels, relief vaults, cavettos and soffits of wooden ceilings. Attention must also be paid to potential defects without visually observable manifestations (cavities and plastered grooves in masonry, different quality of individual masonry components, etc.).

In summary, the most frequently occurring defects and failures of vertical structures of buildings from the second half of the 19th and the start of the 20th centuries include:

- tensile and shear cracks due to the effect of spatial stress states arising in the place of the concentration of compressive stresses in masonry exerted by the mounting of floor beams and girders (or masonry crushing due to the effect of concentrated compression);

- weakening of masonry of e.g., load-bearing pillars by continuous horizontal or vertical grooves, larger openings or the installation of elements differing significantly in stiffness, which contribute to the masonry damage and cracking;
- volume changes due to temperature: unequal or different susceptibility to volume changes caused by the temperature effect can result in different stress states and subsequent failures, e.g., formation of shear cracks, e.g., in the contact between mutually perpendicular masonry walls (perimeter and internal) [34];
- shear forces between the parts of masonry with different temperatures cause microbending deformations and, as a result, can lead e.g., to the failure of the vault mounted on the masonry wall by tensile cracks or damage to the joint between the masonry and the floor structure, etc.;
- sinking of a cantilevered floor structure on which the perimeter masonry of a bay window is mounted; the sinking will cause the appearance of tensile cracks in the bay window masonry whose pattern corresponds to the time pattern of compression trajectories;
- failures of vertical masonry structures built on heterogeneous foundations, in unstable subsoil conditions or in undermined areas is caused by forced deformations, due to the effect of a change in the shape of the footing bottom, i.e., non-uniform subsidence (e.g., due to waterlogging of the foundation joint, e.g., due to leaks in utility networks, waste pipes and incorrect slope of the prepared terrain) or curvature of the footing bottom [35,36].

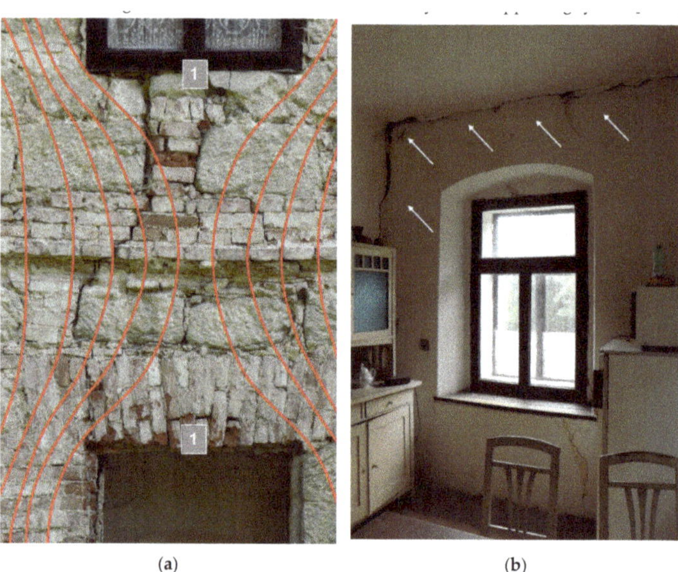

(a)   (b)

Figure 6. Examples of structural failures of structures from the second half of 19th and the start of the 20th century: (**a**) prominent tensile cracks in masonry weakened by openings, with the representation of pressure trajectories, 1—area of normal stresses and tensile cracks; (**b**) crack between the perimeter structure and the ceiling caused by forced deformation due to the effect of a change in the footing bottom.

A change in the shape of the footing bottom and subsequent masonry failures might also have occurred as a result of building activity in the surroundings, e.g., extension of the building, vehicle traffic (dynamic effects causing compaction of the subsoil, shocks of the building), or a change in hydrogeological conditions, waterlogging or frost penetration into the footing bottom of the existing foundation during the erection of a neighbouring building, additional loading of the foundation, dynamic effects and shocks during the construction of a neighbouring building, etc. (Figure 7) [37].

A serious problem of urban masonry apartment buildings is the degradation and disintegration of damp masonry and surface coatings due to the action of moisture contaminated by soluble salts and weak acid solutions. The chemical processes disintegrate the binder component of masonry elements, and the binder and the salts which are the product of these reactions create pressures in the masonry pore system during a change in moisture and degrade its integrity, causing the spalling and disintegration of coatings and layers.

The prevailing parts of historic buildings from the second half of the 19th and the start of the 20th century suffer from increased moisture levels in the foundation masonry and masonry above the foundation line as a consequence of non-functional or damaged waterproofing of the masonry above the foundation line and the masonry of underground parts of the building, leaks or accidents of water supply and sewerage pipes and mains, poor drainage of the terrain, insufficient ventilation, etc. (masonry moisture in dry state is 1%–3% by weight) [38,39]. In many cases of historic brick buildings, there is no damp proofing, or the damp proofing, ventilation systems, etc. have deteriorated and degraded [40].

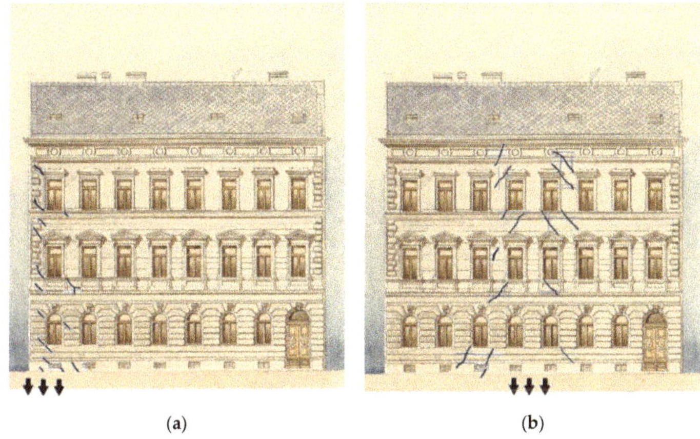

(a) (b)

Figure 7. Graphical representation of the course of cracks and failures of vertical masonry - characteristic tensile cracks on the side (**a**), resp. in the middle (**b**) parts of the building, which arose as a result of the decrease of the affected part and thus the change in the shape of the foundation joint, e.g., an uneven foundation conditions (sands – rock), uneven compaction of the substrate or waterlogging of the soil, etc. (arrows indicate the place of decline of part of the foundations) [41].

The increased moisture levels of underground and partly overground masonry are usually accompanied by chemical and biological degradation processes, whose intensity depends on the value of increased moisture levels and its aggressiveness. The knowledge and assessment of salinity, especially the content of sulphates, chlorides and nitrates in masonry, are essential for the evaluation of the severity of increased moisture levels and the design of an effective remediation method (Figure 8) [42,43]. Salts with hygroscopic properties and the presence of bacteria and mould in the pore system of masonry, which increase the masonry hygroscopicity, can significantly affect the equilibrium moisture content. Degradation processes usually reduce the content of binder components in the masonry. For these reasons, the assessment of the condition of building structures should include, inter alia, an analysis of the change in total porosity and pore distribution and an assessment of the content of water-soluble salts (Figure 9) [44,45]

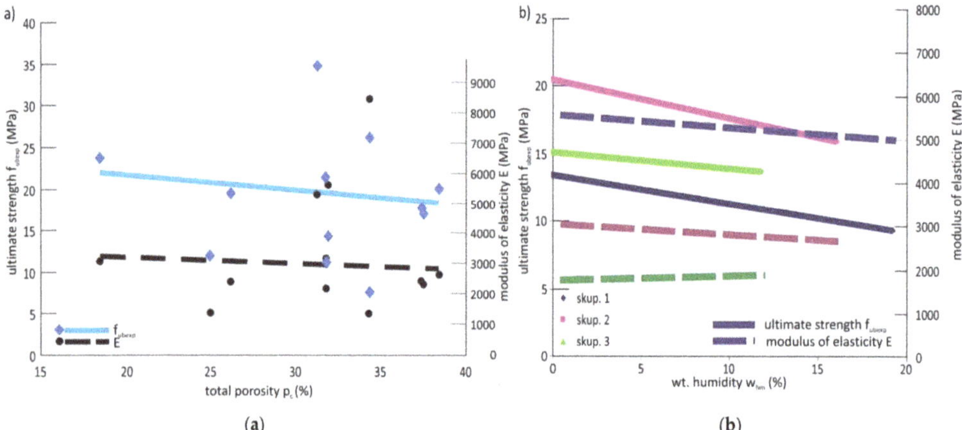

Figure 8. Graphical expression of the dependence of the (**a**) total porosity and modulus of elasticity and ultimate strength, (**b**) wt. moisture and modulus elasticity and ultimate strength of bricks from the end of the 19th century [46].

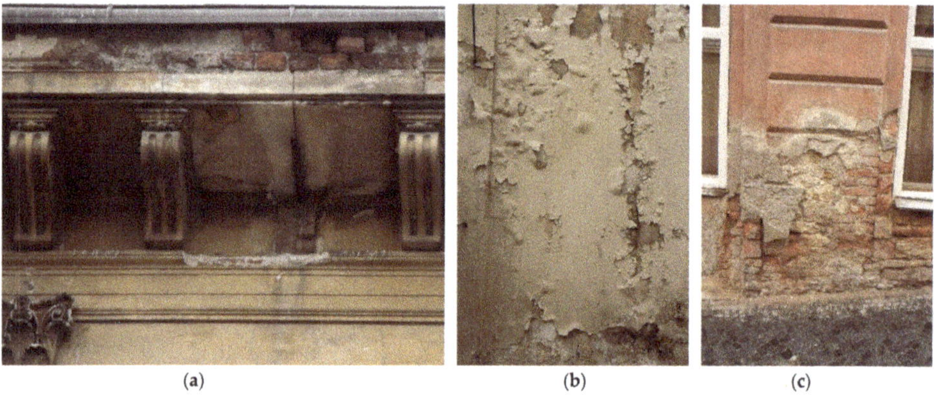

Figure 9. Examples of non-structural failures caused by the effects of the external environment, especially the synergic effect of increased moisture levels and temperature, which are accompanied by chemical, mineralogical and biological processes in structures from the second half of 19th century and the start of 20th century: (**a**) damage of the ledge and its surface finishes, visible salt efflorescence; (**b**) closed and open blisters in the surface layers due to their low diffusion permeability; (**c**) extensive damage to the plinth part caused by the long-term effect of humidity and salts.

3.2. Horizontal Structures

Among the defects of wooden floor structures preceding the occurrence of failures, we might mention, above all, the following: wood infested by wood-destroying fungi or insects, the inappropriate mounting or poor quality workmanship of joints of wooden elements, installation of wood with high moisture levels (whm 12%), impermeable closed-cell wood with initial moisture contents, or closure of high moisture levels in the structure [47].

Other major factors leading to the failures of wooden parts of horizontal load-bearing structures are, among others, the following:

- excessive overall or local loading, wood damaged by unprofessional interventions,
- wood-destroying agents (e.g., due to installation of infested or damp wood without a natural possibility of drying, effect of moisture from leakage and condensation, insufficient ventilation and permeability of the structure) (Figures 10 and 11),

- loosening of joints due to the natural aging and deterioration of major wood properties (water absorption, natural impregnation, elasticity, hardness, strength, toughness),
- change in mechanical properties due to high moisture levels accompanied by the development of wood-destroying processes [48,49],
- thermal bridges in the perimeter masonry in places where wooden beams are mounted in beam pockets, where the wall is weakened by 200 to 250 mm and thus, at the same time, the thermal resistance of these structures is significantly reduced and thermal and moisture conditions of the mounting ends of wooden beams are deteriorated,
- insufficient bending and shear bearing capacity and stiffness of floor beams manifested by their excessive deflection, formation of longitudinal cracks and "slippage" of beams in the places of their mounting and local supports,
- insufficient load-bearing capacity and stiffness of beam floor structures in the horizontal direction (particularly in cases of ineffective or missing wall and beam ties, loosely or insufficiently fixed subflooring to floor beams) can be the cause of a serious threat to structural safety—spatial stiffness—of masonry structures, especially in the case of a building with several storeys, due to horizontal loading effects (eccentricity of vertical loading, non-uniform settlement, wind effects, temperature effects), but mainly dynamic loading effects (caused by technical, induced and natural seismicity); the absence of wall and beam ties, insufficient effectiveness (loosening of joints, insufficient anchoring, etc.) is usually directly manifested by the formation of tensile cracks, damage to the load-bearing masonry and the integrity of the load-bearing system.

(**a**) (**b**)

Figure 10. Examples of failures of wooden parts of horizontal load-bearing structures caused by mounting wood without a natural possibility of drying and with insufficient permeability of the structure: (**a**) rotting head of a wooden ceiling beam; (**b**) yellow patch of wet rot on ceiling wooden panel.

The reserves in the stability of massive longitudinal walls in the transverse direction are insufficient and longitudinal load-bearing walls on their own usually cannot ensure the stability of a building with more than 2 overground storeys in relation to cross wind effects and other effects exerting transverse horizontal forces (e.g., effect of differential settlement, dynamic traffic effects, eccentricity of vertical loads transmitted e.g., by beam floors, temperature differences). This problem is often neglected during rehabilitations when the structural safety (spatial stiffness) may be seriously impaired by the demolition of internal transverse, primarily non-load-bearing walls and partitions.

The reinforcement of a wooden floor structure must be preceded by a detailed survey of wooden structures—particularly a mycological and construction and technical survey, removal of damaged or infested wood and basic chemical protection of wood by a suitable coating, or the sterilization of wood at an elevated temperature [50,51]. The most reliable protection measures of wood against biotic wood-destroying agents are preventive measures against their appearance and development, mainly the protection of wood against increased moisture levels (above 15%).

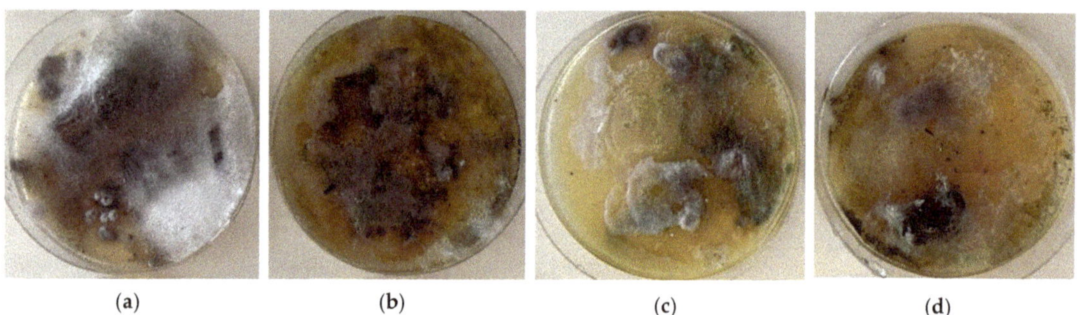

(a)   (b)   (c)   (d)

Figure 11. Example of laboratory evaluation of bioderiogents in cultures of a sample of a damaged ceiling structure in Prague 7—confirmed presence of bacteria and fungi. In all cases, massive development and growth of microscopic fibrous fungi was noted. In cultures of samples were identified: (**a**) *Trichoderma* sp. and bacterias; (**b**) *Trichoderma* sp., *Mucor* sp., wood-destroying fungus and bacterias; (**c**) Mucor sp. and wood-destroying fungus; (**d**) *Penicillium* sp., *Acremonium* sp., *Aspergillus* sp. and bacterias [52].

4. Conclusions

The article presents the most common defects and failures of vertical and horizontal structures of apartment buildings from the second half of the 19th century and the beginning of the 20th century. The construction practice of this period was influenced by three successively issued waves of building regulations, which reflected the building boom of this period.

The issues related to defects and failures, degradation processes and the design of rehabilitation measures of buildings from the second half of the 19th and the start of the 20th century include a wide range of topics of interdisciplinary nature, from the natural sciences, materials engineering, mechanics, elasticity and structural theory of building structures to the knowledge of historical materials, structures and technologies. The knowledge of historical structures, materials and construction methods used helps us to prevent errors in the rehabilitation and restoration of prominent, mainly heritage buildings to the original state.

A precise analysis of the causes and consequences of defects and failures is the basic prerequisite for the reliability and long-term durability of rehabilitation and remediation measures. An integral part of the reconstruction and renovation of buildings must be the elimination of the causes of failures, remediation of all defects that reduce their structural safety, health safety and energy efficiency. To ensure the durability of reconstructed buildings, degradation and corrosive processes caused mainly by chemical and biochemical processes and non-force effects of temperature and moisture must be prevented and adequate protective measures must be taken. In this respect, the protection of buildings and their individual parts against the effects of increased moisture levels is of primary importance.

Funding: This research was funded by the Ministry of Culture of the Czech Republic.

Institutional Review Board Statement: Not applicable.

Informed Consent Statement: Not applicable.

Data Availability Statement: Data available in a publicly accessible repository.

Acknowledgments: This article was written as part of the NAKI DG18P02OVV038 research project "Traditional City Building Engineering and Crafts at the Turn of 19th and 20th Centuries (2018–2022, MK0/DG)".

Conflicts of Interest: The author declare no conflict of interest.

References

1. Janák, P. *Sto Let Obytného Domu Nájemného v Praze*; Styl, Roč. XIII; Styl: Prague, Czech Republic, 1933; p. 63.
2. Dulla, M. *Kapitoly z Historie Bydlení*; CTU in Prague: Prague, Czech Republic, 2014; ISBN 978-80-01-05433-8.
3. Drawing No. 312. Archive of the National Technical Museum: Fund 55—University of Technology Prague, Civil Engineering. Available online: http://www.ntm.cz/en (accessed on 12 January 2021).
4. Ebel, M. *Dějiny Českého Stavebního Práva*; ABF—Arch: Prague, Czech Republic, 2007; ISBN 978-80-86905-21-1.
5. Anzani, A.; Cardani, G.; Condoleo, P.; Garavaglia, E.; Saisi, A.; Tedeschi, C.; Tiraboschi, C.; Valluzzi, M.R. Understanding of historical masonry for conservation approaches: The contribution of Prof. Luigia Binda to research advancement. *Mater. Struct.* **2018**, *51*, 140. [CrossRef]
6. Kroftova, K. (Ed.) *Tradiční Městské Stavitelství a Stavební Řemesla na Přelomu 19. a 20. Století—Svislé a Vodorovné Konstrukce*; CTU in Prague: Praha, Czech Republic, 2020; ISBN 978-80-01-06795-6.
7. Pacold, J. *Konstrukce Pozemního Stavitelství*; Díl II.; Fr. Rivnac Publisher: Prague, Czech Republic, 1890.
8. Niklas, J.; Šanda, F. *J. P. Jöndlovo Poučení o Stavitelství Pozemním*; I. L. Kober Publisher: Prague, Czech Republic, 1865.
9. Archive of the Architecture of the National Technical Museum: Fund 197, No. 20090909/02.
10. Drawing No. 263. Archive of the National Technical Museum: Fund 55—University of Technology Prague, Civil Engineering. Available online: http://www.ntm.cz/en (accessed on 12 January 2021).
11. Maraveas, C.; Wang, Y.; Swailes, T. Fire resistance of 19th century fireproof flooring systems: A sensitivity analysis. *Constr. Build. Mater.* **2014**, *55*, 69–81. [CrossRef]
12. Maraveas, C.; Wang, Y.; Swailes, T. Thermal and mechanical properties of 19th century fireproof flooring systems at elevated temperatures. *Constr. Build. Mater.* **2013**, *48*, 248–264. [CrossRef]
13. Archive of the Architecture of the National Technical Museum: Fund 121—Work of pupils of various vocational schools, No. 20060524.
14. Kohout, J.; Tobek, A. *Konstruktivní Stavitelství*; Part 1; V. Nejedly Publisher: Jaromer, Czech Republic, 1911.
15. Ondřej, S. *Stavba Domu v Praksi*; E. Gregr and Son Publisher: Prague, Czech Republic, 1932.
16. Drawing No. 275. Archive of the National Technical Museum: Fund 55—University of Technology Prague, Civil Engineering. Available online: http://www.ntm.cz/en (accessed on 12 January 2021).
17. Kroftová, K. *Noninvasive Methods of Stabilization and Consolidation of the Historical Surfaces*; Habilitation Thesis; CTU Publishnig: Praha, Czech Republic, 2018. (In Czech)
18. Thomaz, E.; Sousa, H.; Roman, H.R. *Defects in Masonry Walls Guidance on Cracking: Identification, Prevention and Repair*; CIB: Rotterdam, The Netherlands, 2014; ISBN 978-90-6363-090-4.
19. Witzany, J.; Karas, J.; Zlesak, J.; Kupilík, V. Hodnocení vlivu degradačních procesů a historie zatížení na únosnost zděných konstrukcí. In *Sborník Vědeckovýzkumných Grantových Úkolů v Roce 1994*; CTU in Prague: Prague, Czech Republic, 1995; pp. 38–39.
20. Drdácký, M.; Slizkova, Z.; Valach, J. (Eds.) *Příspěvek Technických Věd k Záchraně a Restaurování Památek*; ÚTAM AV ČR: Prague, Czech Republic, 2015. [CrossRef]
21. Domingues, J.C.; Ferreira, T.M.; Vicente, R.; Negrão, J.H. Mechanical and Typological Characterization of Traditional Stone Masonry Walls in Old Urban Centres: A Case Study in Viseu, Portugal. *Buildings* **2019**, *9*, 18. [CrossRef]
22. Cooper, A.J. Structural Failure Patterns of Mid-Nineteenth Century Masonry Buildings of Charleston. Master's Thesis, Clemson University, Charleston, CA, USA, 2018. Available online: https://tigerprints.clemson.edu/all_theses/2830 (accessed on 27 December 2020).
23. Binda, L.; Saisi, A.; Tiraboschi, C. Investigation procedures for the diagnosis of historic masonries. *Constr. Build. Mater.* **2000**, *14*, 199–233. [CrossRef]
24. Witzany, J.; Čejka, T.; Sýkora, M.; Holický, M. Assessment of compressive strength of historic mixed masonry. *J. Civ. Eng. Manag.* **2015**, *22*, 391–400. [CrossRef]
25. Girsa, V.; Michoinova, D. *Historicke Omitky—Zachrana, Konzervace, Obnova*; CTU in Prague: Prague, Czech Republic, 2013; ISBN 978-80-01-05229-7.
26. Witzany, J.; Zigler, R.; Čejka, T.; Pospíšil, P.; Holický, M.; Kubát, J.; Maroušková, A.; Kroftová, K. Physical and Mechanical Characteristics of Building Materials of Historic Buildings. *Civ. Eng. J.* **2017**, *4*, 343–360. [CrossRef]
27. Matysek, P.; Witkowski, M. A Comparative Study on the Compressive Strength of Bricks from Different Historical Periods. *Int. J. Arch. Herit.* **2016**, *10*, 396–405. [CrossRef]

28. Research MSM6840770001. *Spolehlivost, Optimalizace a Trvanlivost Stavebních Materiálů a Konstrukcí*; Faculty of Civil Engineering, CTU in Prague: Praha, Czech Republic, 2010.
29. Kotlík, P.; Šrámek, J.; Kaše, J. *Monografie STOP (2000): Opuka*; STOP: Praha, Czech Republic, 2000.
30. Rovnaníková, P. *Monografie STOP: Omítky—Chemické a Technologické Vlastnosti*; STOP: Praha, Czech Republic, 2002.
31. Zigler, R. Rekonstrukce a sanace nosných zděných kosntrukcí (mechanismus porušování tlařených zděných konstrukcí). Ph.D. Thesis, ČVUT, Fakulta stavební, Katedra konstrukcí pozemních staveb, Praha, Czech Republic, 19 June 2006.
32. Zigler, R. Analýza napjatosti, přetváření a porušování tlačených zděných prvků (stěna, pilíř, valená klenba) se zaměřením na vliv účinku vlhkosti. Master's Thesis, ČVUT, Fakulta stavební, Katedra konstrukcí pozemních staveb, Praha, Czech Republic, 21 January 1999.
33. Witzany, J.; Wasserbauer, R.; Cejka, T.; Kroftova, K.; Zigler, R. *Obnova a Rekonstrukce Staveb. Poruchy, Degradace, Sanace*; CTU in Prague: Prague, Czech Republic, 2018; ISBN 978-80-01-06360-6.
34. Beran, P. The impact of the masonry temperature during restoration to the thermal stress of historic masonry. In Proceedings of the Fifteenth International Conference on Civil, Structural and Environmental Engineering Computing, Civil-Comp Proceedings, Praha, Czech Republic, 1–4 September 2015; Kruis, J., Tsompanakis, Y., Topping, B.H.V., Eds.; Kippen: Civil-Comp Press: Stirlingshire, UK, 2015; Volume 108. [CrossRef]
35. Roca, P.; Lourenco, P.B.; Gaetani, A. *Historic Construction and Conservation—Materials, systems and Damage*; Routledge: Abingdon-on-Thames, UK, 2019; ISBN 9780429052767. [CrossRef]
36. Kuklík, P.; Brouček, M. *Vliv Hydraulických Gradientů a Kolísání Hladiny Podzemních Vod na Stabilitu Staveb*; Zpravodaj ČSVTS: Prague, Czech Republic, 2017; Volume 43, pp. 48–51.
37. Bradáč, J. *Základové Konstrukce*; Akademické Nakladatelství CERM: Brno, Czech Republic, 1995.
38. Witzany, J.; Zigler, R.; Čejka, T.; Maroušková, A.; Kubát, J. Research into the effect of grouting on physical-mechanical properties of historic masonry. In *Advances in Engineering Materials Structures and Systems: Innovations, Mechanics and Applications*; Taylor & Francis: London, UK, 2019; pp. 617–618. ISBN 978-1-138-38696-9.
39. Witzany, J.; Hruška, A.; Zlesák, J.; Zigler, R. Vliv vlhkosti na redistribuci napětí tlačených zděných konstrukcí. In *11. Mezinárodní Vědecká Konference VUT Brno*; VUT v Brně: Brno, Czech Republic, 1999; pp. 239–242.
40. Skabrada, J. *Konstrukce Historickych Staveb*; Argo: Prague, Czech Republic, 2003; ISBN 80-7203-548-7.
41. Ottosen, L.M.; Pedersen, A.J.; Rörig-Dalgaard, I. Salt-related problems in brick masonry and electrokinetic removal of salts. *J. Build. Apprais.* **2007**, *3*, 181–194. [CrossRef]
42. Hendrickx, R. Using the Karsten tube to estimate water transport parameters of porous building materials. *Mater. Struct.* **2013**, *46*, 1309–1320. [CrossRef]
43. Fára, P. *Monografie STOP: Sanace Vlhkého Zdiva*; STOP: Prague, Czech Republic, 2003.
44. Drawing No. 307. Archive of the National Technical Museum: Fund 55—University of Technology Prague, Civil Engineering. Available online: http://www.ntm.cz/en (accessed on 12 January 2021).
45. Delgado, J.M.P.Q.; Guimarães, A.S.; De Freitas, V.P.; Antepara, I.; Kočí, V.; Černý, R. Salt Damage and Rising Damp Treatment in Building Structures. *Adv. Mater. Sci. Eng.* **2016**, *2016*, 1–13. [CrossRef]
46. Cejka, T. *Experimentální Výzkum Vlivu Vlhkosti na Mechanické Vlastnosti Historických Zděných Konstrukcí*; Habilitation Thesis; CTU Publishnig: Praha, Czech Republic, 2009. (In Czech)
47. Reinprecht, L.; Štefko, J. *Dřevěné Stropy a Krovy*; ABF publisher: Prague, Czech Republic, 2000; ISBN 80-86165-29-9.
48. Šimůnková, E.; Kučerová, I. *Monografie STOP: Dřevo*; STOP: Prague, Czech Republic, 2008.
49. Ryparová, P.; Wasserbauer, R.; Rácová, Z. The Cause of Occurrence of Microorganisms in Civil Engineering and the Dangers Associated with their Growth. *Procedia Eng.* **2016**, *151*, 300–305. [CrossRef]
50. Kloiber, M.; Drdácký, M. *Diagnostika Dřevěných Konstrukcí*; ČKAIT: Praha, Czech Republic, 2015; ISBN 978-80-87438-64-0.
51. Drdácký, M.; Kloiber, M. In Situ Compression Stress-Deformation Measurements along the Timber Depth Profile. *Adv. Mater. Res.* **2013**, *778*, 209–216. [CrossRef]
52. Racová, Z.; Loušová, I.; Cejka, T. *Protocol No.124029/2019 about the Test: Survey of the Roof of the Tenement House on Poupetova*; Prague; CTU: Praha, Czech Republic, 2019.

Article

Critical Analysis about Emerging Technologies for Building's Façade Inspection

Ilídio S. Dias [1,*], Inês Flores-Colen [1,2] and Ana Silva [1]

[1] CERIS, Instituto Superior Técnico, Universidade de Lisboa, Avenida Rovisco Pais, 1049-001 Lisbon, Portugal; ines.flores.colen@tecnico.ulisboa.pt (I.F.-C.); anasilva931@msn.com (A.S.)

[2] Department of Civil Engineering, Architecture and Georesources, Instituto Superior Técnico, Universidade de Lisboa, Avenida Rovisco Pais, 1049-001 Lisbon, Portugal

* Correspondence: ilidio.dias@tecnico.ulisboa.pt

Abstract: The diagnosis of the building's façades pathology is extremely important to support rational and technically informed decisions regarding maintenance and rehabilitation actions. With a reliable diagnosis, the probable causes of the anomalies can be correctly identified, and the correction measures adopted can be more compatible with the existing elements, promoting the durability of the façades. Visual inspection is the most common approach to identify anomalies in a building's façade and, in many cases, this technique is sufficient to support the decision to intervene. However, the pathological phenomenon is complex, and the anomalies observed may indicate the presence of other defects, or some anomalies may not be visible in a simple visual observation. This study intends to discuss the application of emerging technologies on the diagnosis and anamneses of building's façade, in order to automatise the collection of reliable on-site data and, thus, reduce the uncertainty of the diagnosis. The use of these techniques can help existing inspection methodologies, already tested, based mainly on the visual assessment of the buildings' elements degradation condition.

Keywords: diagnosis; advanced technologies; façade; claddings; building inspections

Citation: Dias, I.S.; Flores-Colen, I.; Silva, A. Critical Analysis about Emerging Technologies for Building's Façade Inspection. *Buildings* **2021**, *11*, 53. https://doi.org/10.3390/buildings11020053

Academic Editor: Alessandra Aprile

Received: 13 January 2021
Accepted: 1 February 2021
Published: 4 February 2021

Publisher's Note: MDPI stays neutral with regard to jurisdictional claims in published maps and institutional affiliations.

1. Introduction

In recent years, the international scientific community has gained more attention in the field of inspection, diagnosis, maintenance, and rehabilitation of buildings. The maintenance and rehabilitation are considered key factors in buildings' sustainability since these interventions increase the service life of buildings [1]. Inspection and diagnosis of building elements are a very important task to support decision and efficiency of building façade maintenance from the building facades [2,3]. However, there is some uncertainty during these procedures associated with the observation of the degradation phenomenon and the diagnosis process [4].

Collecting information from fieldwork about natural degradation of building elements, is an alternative to laboratory tests like artificial weathering cycles. The fieldwork survey could be very useful to compare with the laboratory tests, since it allows identifying the real-life effects of weathering on the performance of building elements, considering the simultaneous effects of different climatic degradation agents [5–8]. In facades, the degradation condition could be assessed from visual inspections, which involve a non-destructive inspection method. This type of analysis is a natural way to judge the service life of these constructive elements since the maintenance decisions are made based on this type of assessment [9,10]. In that context, several works have been developed, concerning the degradation of buildings and their elements, based on visual inspections [11–14]. These studies can be complemented by some expedient tools, like binoculars to get a closer observation, a tape measure to obtain some dimensional information, a crack with a ruler, and a colour system sample [2]. Other additional diagnosis techniques can also be applied, whether they are non-destructive or destructive techniques. However, visual inspections intend to be easily applicable in practice and without high costs.

Visual inspection is the first method used to evaluate the condition of many infrastructures and buildings [15–17]. Despite the limitations of visual inspections, this method can immediately provide general information about the condition of the building elements [9,18]. This task is usually done by an expert, which performs a survey looking for the damage in order to determine the areas that need intervention. The results of that visual inspection are reliable when dealing with easily visible parts of the façade. However, usually it is not easy, for a surveyor, to analyse the top of the façade or assess anomalies that are in deeper locations without appropriate means of access, and with unfavourable weather conditions. In that context, a reliable and rapid assessment of the defect areas on a large façade may be a difficult work for the surveyor [4,19–23]. Moreover, a visual inspection is highly dependent on the surveyor's expertise [2,24]. Hence, a visual assessment can be considered subjective, human-dependent, time-consuming, and can have low accuracy in defects' measurements in certain situations [16,24]. However, based on the knowledge regarding the condition of the building's elements, is crucial for the adoption of adequate maintenance strategies and it is fundamental for life cycle analysis, considering the economic and environmental costs of the building components over their life cycle [23,25,26].

The limitations of visual inspections motivated the use of systemized alternatives such as remote sensing for mapping the defective areas on a surface of building façades [19,22]. In recent years, the interest about advanced technologies to automate the inspection of building façades has emerged. Some studies have been developed in the field of building's façade inspection, mainly in cultural heritage buildings, by focusing on building an energy audit [21,23,27–31]. Novel technologies for a reality capture have been applied, mainly terrestrial laser scanning, photogrammetry, infrared thermography, photogrammetry, and drones. The application of these technologies can improve the traditional survey, producing more objective results in a faster way [23,32]. They could automate and improve the visual inspection, reducing the subjectivity associated with the inspector's survey [17,24]. Nevertheless, automation as a way to detect and measure a given defect is still a great challenge, which needs several technologies combined for accurately mapping all the anomalies that can occur in a given building component. More research is needed to put into practice these technologies to support the facades' inspections.

Most of these studies only focus on a single technology. Different defects could require different technologies to support and automate building's facades' diagnosis. In this sense, this study intends to discuss the suitability of these advanced technologies to automate a visual assessment during field inspection and the new trends for inspection of buildings' facades with a focus in current buildings. Furthermore, the most suitable technology to support the diagnosis of different kinds of defects is identified. These technologies can bring some automation for the inspection and diagnosis of buildings and their components, reducing the inspection time in the field, avoiding several visits to the building's location (because all the necessary information is not collected during the first time), and producing more reliable information about the anomalies present in the component under analysis.

2. Materials and Methods

In order to discuss the most suitable technologies to automate a visual assessment of several defects in buildings' facades, this research follows a specific methodology. To better understand the needs of advanced technologies, two case studies are presented, to illustrate the application of a building inspection system, based on a visual assessment of the degrading building components. In this inspection system, the parameters involved in the calculation of the facades' degradation condition are explained in detail and some improvements are suggested. In addition, to illustrate some uncertainty associated with the visual inspection, the same 47 natural stone claddings are inspected by two surveyors in different moments. Then, an extensive literature review, with about 50 references, is performed to systematise the technologies able to automate the visual façade inspection in buildings. In the end, the suitability of previous technologies is presented, in the sense of an automated visual assessment, according to different defects present in facades. Some recommendations are made in this topic. In the next sections, these topics will be developed and discussed in detail.

3. Global Inspection System for the Buildings' Envelope Elements: 2 Case Studies

The adoption of adequate maintenance strategies is the most efficient way of increasing the durability of buildings and their elements. However, an adequate intervention needs an accurate and reliable diagnosis of the defects and possible causes. Each building is unique and presents different types of defects. Nevertheless, certain patterns can be identified when analysing a large sample of buildings. A systematic analysis of the data collected during the building inspections can provide a reliable database for a guidance to prevent the occurrence of defects and to repair the existing ones. For these reasons, several inspection systems have been developed by different authors. Ferraz [1] presented and discussed 10 building pathology databases, developed between 1982 and 2013. More recently, Lee [33] proposed an inspection methodology to introduce information about building defects into BIM (Building Information Model). Bortolini and Forcada [26] proposed a building inspection system for the assessment of whole building performance [34]. Gonçalves [35] analysed the existing methods of inspection and diagnosis for ancient buildings, based on visual inspections.

In fact, it is almost consensual that building inspection benefits from the standardisation of inspection and diagnosis procedures [36,37]. The adoption of inspection systems helps surveyors to have harmonised procedures, which makes inspection reports more consistent and simpler to understand [2,3,38].

In recent years, different methodologies for the inspection and diagnosis of non-structural elements of the building envelope have been established, in order to create a reliable database and systemise procedures for the on-site assessment of defects [39–41]. Moreover, several studies in service life prediction methodologies have been developed, with the aim to evaluate when it is necessary to intervene based on the element's degradation condition assessment [42,43].

Under the SLP for BMS (Service Life Prediction for a risk-based Building Management System) research project, a global inspection system was developed [44]. This system embraces 12 building elements, namely: external claddings of pitched roofs, flat roofs, adhesive ceramic tiling, natural stone claddings, wood floorings, door and window frames, epoxy resin floorings, vinyl and linoleum floorings, wall renders, external thermal insulation composite systems (ETICS), painted façades, and architectural concrete surfaces. The unification of these partial inspection systems implied the harmonisation of the classification of defects, probable causes, diagnosis methods, and repair techniques in a single system. This global inspection system intends to provide a solid basis for surveyors to inspect the building envelope through standardisation of procedures [3,34].

This global system is based on visual and physical scales to characterize the building envelope elements during the inspections. Gaspar & de Brito [45] begin this work with the development of a qualitative scale, based on the assessment of the physical and visual degradation of rendered façades, analysed during a fieldwork survey, which can be associated with a quantitative index. This describes the global performance of the façades [42]. This numerical index, called severity of degradation (S_w), expresses the global degradation of a given façade through the ratio between a weighted degraded area and a reference area, equivalent to the total cladding with the highest possible level of degradation (Equation (1)). The weighted degraded area is a product of the façade area affected by different groups of anomalies with a weighted factor related to the severity of each anomaly detected (k_n)

and a weighted factor associated with relative weight of each anomaly on the overall degradation of façade ($k_{a,n}$).

$$S_w = \frac{\sum(A_n \times k_n \times k_{a,n})}{A \times \sum(k_{max})} \tag{1}$$

where S_w is a severity of degradation of the cladding, expressed as a percentage. k_n is the multiplying factor of anomaly n, as a function of their degradation level, within the range $K = \{0, 1, 2, 3, 4\}$. $k_{a,n}$ is a weighting factor corresponding to the relative weight of the anomaly detected. An is the area of cladding affected by an anomaly n. A is the total area of the cladding, and k_{max} is the multiplying factor corresponding to the highest degradation level of the cladding [42]. This methodology was developed to external renders and applied to the building elements previously mentioned.

To discuss the reliability of this severity of the degradation (S_w) index, two case studies are presented in this paper: (i) a continuous element (a rendered wall) and (ii) a discontinuous element (a natural stone cladding). These two claddings represent different challenges in the calculation of severity of the degradation index, which will be presented later. Figure 1 describes a rendered wall inspected according to the global inspection system. The case study is a building located in Tavira, Algarve (South of Portugal). The façade is oriented toward the west and present an average level of protection against the wind and rain. In terms of environmental factors, the inspected building is located at less than 3 km from the sea in a rural environment and exposed to normal conditions of air temperature and relative humidity. The total area of the facade analysed is 105.52 m^2 and the render is 10 years old.

Figure 1. First case study: rendered façade (data sourced from Reference [46]).

Table 1 presents the defects in the rendered façade (Figure 1) observed during the visual inspection carried out during the fieldwork survey. The surveyor recorded the area of rendering affected by each group of defects (A_n). The degradation condition level of each defect found (k_n), according to its condition (taking into account the classification of defects proposed in Table 2), and the factor corresponding to the relative weight of each defect in the overall degradation of the façade ($k_{a,n}$), according to Table 3. In summary, Table 1 presents the weighted degraded area ($A_n \times k_n \times k_{a,n}$), taking into account the defects observed. In order to calculate the severity of the degradation (S_w) index of this cladding, Equations (2) and (3) are used.

Table 1. Identification of anomalies observed in a rendered façade, after an inspection.

Defects		A_n (%)	A_n (m^2)	k_n(-)	$k_{a,n}$ (-)	$A_n \times k_n \times k_{a,n}$ (m^2)
Stains, A_{stains}	Condition B	2.5	2.63	1	0.67	1.762
	Condition C	4.2	4.41	2	0.67	5.909
	Condition D	3.1	3.26	3	0.67	6.553
	Condition E	-	-	4	0.67	-
Cracking, $A_{cracking}$	Condition B	-	-	1	1.00	-
	Condition C	3.3	3.47	2	1.00	6.940
	Condition D	1.8	1.91	3	1.00	5.730
	Condition E	4.1	4.30	4	1.50	25.800
Loss of adhesion, $A_{adhesion}$	Condition B	-	-	1	1.50	-
	Condition C	-	-	2	1.50	-
	Condition D	0.8	0.85	3	1.50	3.825
	Condition E	-	-	4	1.50	-

Table 2. Classification of the degradation condition of rendered facades to the level of the defect (k_n) [1].

Degradation Condition	Physical and Visual Assessment	Illustrative Example
Condition A (k_n = 0)	Complete mortar surface with no deterioration, with surface even and uniform. No visible cracking or cracking $\leq$ 0.1 mm. Uniform colour and no dirt. No detachment of elements.	
Condition B (k_n = 1)	Non-uniform mortar surface with capillary cracking (0.1 to 0.25 mm). Slight stains in localized areas, mainly dirt.	
Condition C (k_n = 2)	Non-uniform mortar surface with likelihood of hollow localized areas determined by percussion, but no signs of detachment. Small cracking (0.25 to 1.0 mm) in localized areas. Changes in the general colour of the surface with a potential presence of microorganisms.	
Condition D (k_n = 3)	Mortar with localized detachments or perforations, revealing a hollow sound by percussion. Detachments only in the socle. Easily visible cracking (1 to 2 mm). Dark stains of damp and dirt, often with microorganisms and algae.	
Condition E (k_n = 4)	Incomplete mortar surface due to detachments and falling of mortar patches. Wide or extensive cracking ($\geq$ 2 mm). Very dark stains of damp and dirt, often with microorganisms and algae.	

[1] Data sourced from References [42,47].

Table 3. Weighted factor for render facades ($k_{a,n}$) [1].

Degradation Condition	Stains	Cracking	Loss of Adhesion
Condition A	0.00	0.00	0.00
Condition B	0.67	1.00	1.50
Condition C	0.67	1.00	1.50
Condition D	0.67	1.00	1.50
Condition E	0.67 *	1.50	1.50

* 1.00 in the situations of occurrence of ice / thaw cycle. [1] Data sourced from References [42,47].

In Equation (3), the calculation of the severity of degradation (S_w) for this case study is presented. The result is a S_w index of 13.39%, which means a slight degradation of rendered façade, according to Table 4.

$$S_w = \frac{\sum(A_{stains} \times k_n \times k_{a,n}) + \sum\left(A_{cracking} \times k_n \times k_{a,n}\right) + \sum(A_{adhesion} \times k_n \times k_{a,n})}{A \times k_{max.}} \tag{2}$$

$$S_w = \frac{\overbrace{1.762 + 5.909 + 6.553}^{A_{stains}} + \overbrace{6.940 + 5.730 + 25.800}^{A_{cracking}} + \overbrace{3.825}^{A_{adhesion}}}{105.52 \times 4} \times 100 = \frac{56.519}{422.08} \times 100 = 13.39\% \tag{3}$$

Table 4. Correspondence between the degradation condition and severity of degradation of rendered facades [1].

Degradation Condition	Severity of Degradation, S_w
Condition A (no degradation)	$S_w \leq 1\%$
Condition B (good)	$1\% < S_w \leq 5\%$
Condition C (slight degradation)	$5\% < S_w \leq 15\%$
Condition D (moderate degradation)	$15\% < S_w \leq 30\%$
Condition E (generalized degradation)	$S_w \geq 30\%$

[1] data sourced from References [42,47].

Figure 2 describes another case study, a natural stone cladding, directly adhered to the substrate, which was inspected using the global inspection system. This case study refers to a building located in Parque das Nações, Lisbon (Centre of Portugal). The façade is oriented toward the east and presents an average level of protection against the combined action of wind and rain. In terms of environmental factors, the inspected building is located at less than 3 km from the sea, in an urban environment, and exposed to unfavourable conditions of air temperature and relative humidity. The total area of the facade analysed is 75 m^2 and the natural stone cladding is 12 years old.

The defects observed in the natural stone cladding inspected (Figure 2) are synthetized in Table 5. The surveyor recorded the area of natural stone affected by each group of the defect (A_n): A_v—visual anomalies, A_j—joint anomalies, A_f—bond-to-substrate anomalies, and A_i—loss-of-integrity anomalies. The degradation condition level of each defect found (k_n) and the factor corresponding to the relative weight of each defect on the overall degradation of façade ($k_{a,n}$) are assigned according to the condition classification in Table 6. In summary, Table 5 presents the weighted degraded area ($A_n \times k_n \times k_{a,n}$) in order to calculate the severity of the degradation (S_w) index (Equations (4) and (5)).

Figure 2. Second case study: natural stone cladding.

Table 5. Identification of anomalies observed in natural stone cladding after performing façade inspection.

Defects		A_n (%)	A_n (m^2)	k_n (-)	$k_{a,n}$ (-)	$A_n \times k_n \times k_{a,n}$ (m^2)
Visual, A_v	Condition B	-	-	1	0.13	-
	Condition C	40.3	30.24	2	0.13	7.862
	Condition D	-	-	1	0.13	-
	Condition E	-	-	1	0.13	-
In joints, A_j	Condition B	-	-	1	-	-
	Condition C *	20	15	2	0.25	7.500
	Condition C **	10	7.50	2	1	15.000
	Condition D	-	-	3	-	-
	Condition E	-	-	4	-	-
Bound-to-substrate, A_f	Condition B	-	-	1	1.20	-
	Condition C	10	7.50	2	1.20	18.000
	Condition D	-	-	3	1.20	-
	Condition E	-	-	4	1.20	-
Loss-of-integrity, A_i	Condition B	-	-	1	1	-
	Condition C	10.8	8.10	2	1	16.200
	Condition D	20	15	3	1	45.000
	Condition E	-	-	4	1	-

* joint anomalies only with material degradation. ** joint anomalies with material loss—open joint.

Table 6. Classification system for natural stone claddings [1].

Degradation Condition	Anomalies		$k_{a,n}$	% Area of NSC Affected	Severity of Degradation (%)
A (k_n = 0)	No visible degradation		-	-	$S_w \leq 1$
B (k_n = 1)	Visual or surface degradation anomalies	Surface dirt	0.13	>10	$1 < S_w \leq 8$
		Moisture stains/localised stains/colour change	0.13	≤ 15	
		Flatness deficiencies	0.13	≤ 10	
	Loss-of-integrity anomalies	Material degradation * $\leq$ 1% plate thickness	1.00	-	
		Material degradation * $\leq$ 10% plate thickness Cracking width $\leq$ 1 mm	1.00	≤ 20	

Table 6. *Cont.*

Degradation Condition		Anomalies	$k_{a,n}$	% Area of NSC Affected	Severity of Degradation (%)
C (k_n = 2)	Visual or surface degradation anomalies	Moisture stains/localised stains/colour change	0.13	>15	8 < S_w ≤ 20
		Moss, lichen, algae growth/parasitic vegetation/efflorescence	0.13	≤30	
		Flatness deficiencies	0.13	>10 and ≤50	
	Joint anomalies	Joint material degradation	0.25	≤30	
		Material loss-open joint	1.00	≤10	
	Bond-to-substrate anomalies	Scaling of stone near the edges Partial loss of stone material	1.20	≤20	
	Loss-of-integrity anomalies	Material degradation * ≤ 10% plate thickness Cracking width ≤ 1 mm	1.00	>20	
		Material degradation * > 10% and ≤ 30% plate thickness Cracking width > 1 mm and ≤ 5 mm	1.00	≤20	
		Fracture	1.00	≤5	
D (k_n = 3)	Visual or surface degradation anomalies	Moss, lichen, algae growth/parasitic vegetation/efflorescence	0.13	>30	20 < S_w ≤ 45
		Flatness deficiencies	0.13	>50	
	Joint anomalies	Joint material degradation	0.25	>30	
		Material loss-open joint	1.00	>10	
	Bond-to-substrate anomalies	Scaling of stone near the edges Partial loss of stone material	1.20	>20	
		Loss of adherence	1.20	≤10	
	Loss-of-integrity anomalies	Material degradation * > 10% and ≤ 30% plate thicknessCracking width > 1 mm and ≤ 5 mm	1.00	>20	
		Material degradation * > 30% plate thickness Cracking width > 5 mm	1.00	≤20	
		Fracture	1.00	>5 and ≤10	
E (k_n = 4)	Bond-to-substrate anomalies	Loss of adherence	1.20	>10	S_w > 45
	Loss-of-integrity anomalies	Material degradation * > 30% plate thickness Cracking width > 5 mm	1.00	>20	
		Fracture	1.00	>10	

* Material degradation is meant to be every anomaly that involves loss of volume of the stone material. [1] data sourced from Reference [42].

In Equation (5), the calculation of the severity of degradation (S_w) of the second case study is presented. The result is a S_w index of 10.43%, which means a slight degradation (Condition C) of the natural stone cladding analysed, according to Table 7.

$$S_w = \frac{\sum(A_v \times k_n \times k_{a,n}) + \sum(A_j \times k_n \times k_{a,n}) + \sum\left(A_f \times k_n \times k_{a,n}\right) + \sum(A_i \times k_n \times k_{a,n})}{A \times k_{max}.}$$

(4)

$$S_w = \frac{\overbrace{7.862}^{A_v} + \overbrace{7.5 + 15}^{A_j} + \overbrace{18}^{A_f} + \overbrace{16.2 + 45}^{A_i}}{75 \times 14} \times 100 = \frac{109.562}{1050} \times 100 = 10.43\% \tag{5}$$

Table 7. Correspondence between the degradation condition and the severity of degradation of natural stone claddings [1].

Degradation Condition	Severity of Degradation, S_w
Condition A (no degradation)	$S_w \leq 1\%$
Condition B (good)	$1\% < S_w \leq 8\%$
Condition C (slight degradation)	$8\% < S_w \leq 20\%$
Condition D (moderate degradation)	$20\% < S_w \leq 45\%$
Condition E (generalized degradation)	$S_w \geq 45\%$

[1] data sourced from Reference [48].

In these two case studies, the visual inspection was aided by a crack ruler to measure the thickness of the cracks and a tape measure to get dimensions to support the quantification of defected areas (A_n) in the stone cladding and rendered façades. These procedures strongly influence the parameters used to calculate the severity of the degradation (S_w) index. Furthermore, these inspections are a time-consuming process. Only the natural stone cladding in the bottom wall was analysed, and, in some areas, a ladder was needed to evaluate the defects present in the façade in more detail at a higher level. These examples confirm that the reliability of this global inspection system depends on the accuracy of the data collection, and it is intrinsically related to the surveyor's expertise and the inspection conditions. Some automation in data collection could help to obtain more reliable and standardised results and reduce the acquisition time of the data.

4. The Uncertainty Associated with the Building Inspection Based on Fieldwork

The global inspection system adopted for the inspection of the facades previously analysed, is based only on the visual assessment of the components, thus, encompassing some uncertainty on the quantification of the degraded areas, as discussed previously. To illustrate this issue, Table 8 shows the results of the inspections carried out on the same 47 natural stone claddings by two surveyors. The two surveyors perform these inspections as part of their masters' thesis in civil engineering [40,48]. After processing the collected data acquired by different methodologies, the two inspectors obtained different results in some façades, leading to a variation between 0% to 2.23% in the severity of the degradation index—S_w (variations %), as shown in Table 8. This could be related to some differences in the way data is collected by the surveyor, since they have different goals, which results in slightly different approaches, and the subjectivity associated with visual inspection is due to the assessment of each surveyor. Neto developed her work in the inspection and diagnosis of natural stone cladding [40] and Silva in a field of service life prediction of natural stone cladding [48]. In this sense, Neto [40] only identified the anomalies observed, and did not estimate all the areas affected by each anomaly, while Silva [48] estimated the areas affected by each defect in order to obtain the severity of the degradation index—S_w.

Table 8. The results obtained by two surveyors for the same 47 case studies of facades with natural stone cladding [1].

ID (from Neto, 2008)	S_w (Neto, 2008)	S_w (Silva, 2009)	S_w (Neto, 2008)/ S_w (Silva, 2009)	ΔS_w
Ed. A.1	2.1%	2.1%	1.00	0.00%
Ed. Q.2	1.8%	1.8%	1.00	0.00%
Ed. R.3	1.9%	1.9%	1.00	0.00%
Ed. T.2	2.1%	2.1%	1.00	0.00%
Ed. BB.4	0.6%	0.6%	1.00	0.00%
Ed. BB.5	1.1%	1.1%	1.00	0.00%
Ed. I.1	2.0%	2.0%	0.98	0.03%
Ed. H.3	1.6%	1.5%	1.03	0.05%
Ed. E.2	2.1%	2.0%	1.05	0.10%
Ed. F.2	1.1%	1.2%	0.87	0.16%
Ed. E.1	2.1%	1.9%	1.11	0.21%
Ed. I.3	1.3%	1.1%	1.20	0.21%
Ed. R.2	2.1%	1.9%	1.11	0.21%
Ed. F.1	1.1%	1.4%	0.79	0.29%
Ed. P.2	2.0%	1.7%	1.19	0.32%
Ed. Q.1	1.7%	1.4%	1.23	0.32%
Ed. BB.6	2.4%	2.7%	0.88	0.32%
Ed. I.2	4.3%	3.9%	1.10	0.40%
Ed. R.1	2.1%	1.7%	1.23	0.40%
Ed. Z.2	1.8%	2.3%	0.81	0.43%
Ed. Z.1	1.1%	1.5%	0.71	0.43%
Ed. T.3	0.8%	1.3%	0.63	0.47%
Ed. Z.3	2.1%	2.7%	0.80	0.54%
Ed. O.1	4.1%	3.5%	1.16	0.56%
Ed. K.4	3.0%	2.4%	1.28	0.66%
Ed. H.1	1.2%	0.5%	2.26	0.68%
Ed. T.1	1.5%	2.2%	0.68	0.70%
Ed. G.1	2.1%	2.8%	0.73	0.77%
Ed. D.3	0.9%	0.1%	11.58	0.78%
Ed. S.1	2.1%	3.0%	0.72	0.83%
Ed. K.3	2.0%	2.9%	0.70	0.85%
Ed. Q.3	2.0%	2.8%	0.70	0.86%
Ed. B.1	1.9%	1.0%	2.00	0.96%
Ed. CC.2	1.8%	2.7%	0.64	0.96%
Ed. KK.1	3.4%	2.3%	1.49	1.13%
Ed. AA.3	10.7%	9.5%	1.13	1.21%
Ed. U.1	4.5%	3.3%	1.38	1.23%
Ed. H.2	2.3%	1.0%	2.36	1.31%
Ed. AA.2	10.8%	9.3%	1.16	1.50%
Ed. D.2	2.0%	0.4%	4.85	1.55%
Ed. O.3	0.3%	1.9%	0.17	1.57%
Ed. K.1	3.3%	1.6%	1.99	1.63%
Ed. U.3	3.4%	1.7%	1.98	1.70%
Ed. D.1	2.0%	0.1%	14.22	1.81%
Ed. J.2	3.4%	1.5%	2.29	1.93%
Ed. K.2	2.7%	0.6%	4.20	2.06%
Ed. AA.1	10.7%	8.5%	1.26	2.23%

[1] data sourced from Neto [40] and Silva [48].

The differences between the results are associated with the type of information collected for severity of the degradation index calculation, to the gap in time between inspections and some limitations related to the visual survey, like weather conditions at the inspection moment (e.g., the incidence of solar radiation in the wall can mask some defects). The subjectivity associated with the inspector assessment is related to difficulties in assessment, mapping, and measurements of the defective areas in elevated areas of the building facades. Furthermore, the deviations between the values obtained by the two surveyors seem acceptable, given the subjectivity inherent to the visual inspection. However, other techniques could be applied to automate the inspection of facades, in order to reduce the subjectivity inherent in the visual inspection. Technologies like 3D laser scanning, infrared thermography, photogrammetry, digital image processing, and drones could provide some automation in collecting geometric and photographic data. These techniques intended to overcome some limitations in assessing and measuring the defects areas while improving the evaluation of the degradation condition through the severity of a degradation index (S_w).

5. Emerging Technologies to Automate Visual Building's Façade Inspection

In this section, some advanced technologies to automate a building facade inspection are summarized and discussed. To overcome some limitations in assessing and measuring the defects areas, previously shown in Sections 3 and 4, some technologies were selected to collect geometric and photographic data. The technologies analysed in this study are the most addressed in the literature [16,21,23,24,27–32,49,50]: 3D laser scanning, infrared thermography, photogrammetry, digital image processing, and drones. Each technology is presented succinctly, with the main focus of automate mapping and quantifying the defects observed in building facades. Some advantages and disadvantages are presented, intending to analyse and select the suitability technologies to complement and improve a visual assessment of building facades.

5.1. 3D Laser Scanning (TLS)

Terrestrial laser scanning (TLS) is a process that records a 3D digital image within a determined radius from the location of the laser scanner, through infrared light. The result are 3D point clouds of the objects surface. The use of this technique has grown in cultural heritage and historic preservation projects, mainly because their rapid, wide-ranging, and non-invasive method of documentation, which proved to be cost-effective when compared to traditional techniques. This technique is considered one of the best solutions for 3D digitalization of cultural heritage assets [51–53]. With TLS, it is possible to detect some defects as surface delamination, cracks, displacements, and deflections in walls [20,54]. This technology also allows measuring the defects with the acquired information into a Computer Aided Design (CAD) system [23]. The accuracy and details of the recording depend on the scanner specifications and the distance to the scanned object [23,54]. The mapping technique using a 3D laser scan allows a better prediction of replacement cost of wall surfaces through the location of the defects. Another advantage of this technique is the ability to document large areas at ground level, thus, avoiding the scaffolding costs and creating a safer environment for the surveyors [51].

However, the laser scanner has some drawbacks in collect data from hidden or unreachable surface areas from the ground, like capture points in tall buildings. In a historic urban context, with narrow streets, this technique does not provide good results [55]. Moreover, the equipment and software acquisition can be expensive with equipment prices around 30,000 euros [23,56,57].

However, the use of TLS aids to overcome the complexity in accessing some parts of buildings and the unfavourable lighting conditions, like shaded and lighted areas, because it is independent of solar lighting [23,51,52]. This technology generates coordinates of millions of points in reflecting surfaces, providing a rapid geometric representation of objects [19,23]. It provides a geometrical data with high resolution and accuracy, but usually the radiometric data is not useful to defects mapping, like stains, due to the sensors' limitations [23,29,55]. In that context, several authors [19,55,56] have done research combining TLS with other technologies like digital image processing, photogrammetry, and infrared thermography. Therefore, this approach brings some automation in mapping and measuring some defects in facades, as delamination as well as loss of adhesion and cracks (with some limitation in crack width) in a more accurate and detailed way.

5.2. Infrared Thermography (IRT)

Infrared thermography (IRT) is a non-destructive technique that has been applied in buildings inspection as an important diagnostic tool [37,58]. The principle is based on the measurement of the radiant thermal energy distribution, which is emitted from an object. The thermal energy is measured by an infrared camera [59]. This technology has been used to detected surface defects such as moisture, air leakage in the walls, detachment, and cracks in some type of claddings [37,58,60–65]. Thermography can detect anomalies with surface temperature variations. Methodologies combining infrared thermal images with TLS was performed and showed good results to identify anomalies on masonry walls [66]. Other authors [67,68] combines infrared thermography with photogrammetry techniques, intending to obtain thermographic information where it is possible to measure the defects on the image.

Despite the innumerable capabilities of this technology, there is some limitations related to the significant cost of infrared cameras with high resolution. However, there is a low-cost camera with costs around 500 euros [69]. Thermography is highly dependent on climatic conditions, components of a surrounding natural and built environment, building orientation and shape, surface roughness/texture/colour, and camera settings [58,70]. The façade should not be exposed to wetting, frosting, or direct solar radiation in the acquisition phase. In addition, this technique is significantly dependent on the expertise of the operator [71]. However, this technique could be an upgrade for damage detection in visual inspection with a capability of mapping moisture stains, detachment, loss of adhesion, and cracks, particularly when there is no access to the facade. IRT combined with photogrammetry techniques and TLS could also provide measures of mapped defects.

5.3. Photogrammetry and Remote Sensing (PRS)

The International Society of Photogrammetry and Remote Sensing define photogrammetry and remote sensing (PRS) as "the art, science, and technology of obtaining reliable information from noncontact imaging and other sensor systems about the earth and its environment, and other physical objects and processes through recording, measuring, analysing, and representation" [72]. PRS is a technique capable of determining 3D geometry of physical objects by analysing and measuring 2D photographs. It is divided in aerial and terrestrial photogrammetry. In aerial locations, images are acquired from an aircraft providing topographic maps. In terrestrial photogrammetry, images are acquired near the object and provide dimensional information about the object. In case of the object size and distance camera-to-object of less than 100 m, the technique is defined as close-range photogrammetry [73].

This technique could be applied, through a digital camera, in facades to obtain orthoimages, where, afterward, it is possible to measure the defects [74]. However, several authors have been combining PRS with other technologies, like TLS. With this arrangement, they provide good geometric and radiometric information, so it is possible to measure defects on façade [74,75]. Other authors combine image processing with PRS to automatic crack monitoring [56] and measure defects through the image in building façades [76]. One advantage of PRS is the cost, around 150 euros for the software, assuming there is a computer [77], typically lower than TLS (30,000 euros) [68].

5.4. Digital Image Processing (DIP)

Digital image processing is a technique to extract information from the images with several applications in engineering and architecture. This technique requires the use of software to perform image processing on digital images. There are two main areas of application, which is a low level that involves the improvement of pictorial information for human interpretation and a high level for the processing of scene data for autonomous machine perception, to give the system the ability to interpret and understand an image [78]. More specifically, low-level processing contains the image acquisition, image compression, a pre-processing method for noise filtering, edge extraction, and image sharpening. It also contains a high-level, useful mathematical method, such as neural networks, fuzzy logic, pattern recognition, and artificial intelligence [27,78].

This technique has been implemented in multiple areas such medicine, automation, security, and defence [79]. In recent years, this technique has been adopted in the architecture field for identifying different materials and defects such as in stone masonry facades. This work is based on image processing software, which contains algorithms capable of classifying the stone anomalies [79]. At this stage of image processing, more accuracy is needed. DIP has been used to detect and quantify defects in tiling work [78,80]. Thus, DIP is capable of measuring defective areas and increase the reliability of visual inspection [78], detecting cracks in building facades [81].

However, some image processing limitations are related to image acquisition, the cladding appearance, the camera distance, and position, and with a light condition at the inspection moment. Those factors can affect the image captured and cause some inaccuracies in the defects' detection [78].

5.5. Drones (UAV)

The use and development of drones has its origins in the military field in the middle of the century XIX. Since 2010, the scenery changed with the falling prices and increasing ease operation [70]. The advances in programming and autopilot systems, the miniaturization of components, such as gyros and GPS units, made the machines smaller, cheaper, and easier to fly [49]. Currently, drones or more formally known as Unmanned Aerial Vehicles (UAV), have high popularity and the technology has got a maturity level that makes it more user-friendly and inexpensive. It is possible to purchase a good UAV for prices around 1700 euros [82]. In the construction industry, the use of these devices can contribute to reduce the time of tasks, like monitoring construction activities and inspection of buildings, increase the quality of the work, improve safety standards, and reduce costs [83]. Building pathology and diagnosis can be done with drones faster than conventional methods, more safely, in some circumstances [49]. UAV can use coupled cameras to capture HD images and videos or infrared cameras and 3D laser scanners to identify damages and cracking in building facades [84,85]. UAV can fly in inaccessible areas without risk for the operator, like higher facades in tall buildings, or between buildings in narrow streets. The speed at which the survey can be performed in the field is also an advantage of this technology [49,86].

Some disadvantages are related to the quality of images and videos obtained, which can be influenced by light conditions and inspection distances to elements in some cases. The load of a different type of camera, the meteorological conditions, and physical barriers (e g , trees near the façades) could also be a limitation for this technique [19]. The battery duration remains a big challenge to be solved with the best flight times around 30 min [82]. Additionally, the use of this technology must follow the country regulation [81].

UAV can be used with an infrared thermal camera. As a result of technological advances, the infrared sensors became smaller and lighter, which enabled their use by drones [70]. A geometric and photographic survey is also possible with a drone with a kinetic sensor installed. This solution is characterized as low-cost among others, which is capable of generating 3D models [87]. UAV with coupled cameras are particularly useful in visual inspection and are capable of improving the assessment of some defects, such as cracks and stains.

6. Critical Analysis for Inspection and Diagnosis of Facades Elements

In this section, the suitability of previous technologies presented to automate visual assessment of building facades is discussed. The aim of using these technologies is to collect more reliable data during fieldwork. The key improvement of the global inspection system relies on the calculation of the severity of degradation index (S_w), which evaluates the condition state of facade in real exposure conditions, based on visual inspections, as mentioned in Section 3. Consequently, there are several technologies capable of bringing some automation in the visual inspection, as mentioned in Section 5. However, the measurement of defective areas in facades is still a big challenge, as discussed in this section.

As previously shown in Section 4, the calculation of severity of the degradation index (S_w), in a global inspection system, could be improved if the collected data is more accurate and independent of the surveyor expertise as well as visual inspection conditions. Thus, advanced technologies can help in this context. The selected technique must be able to map the defects and mainly measure with an adequate accuracy (or with a known margin of error) the area affected by each anomaly type (e.g., cracks, detachments, and stains). With that purpose, in Table 9, a critical analysis of emerging technologies (selected in Section 5) is presented to map and measure defect areas in building facades, based on the accuracy of field data collection and based on the importance of this data for the calculation of the severity degradation index (S_w).

Table 9. Critical assessment of emerging technologies to measure defect areas in facades, according to the nature of defects.

		Emerging Technologies				
		TLS	**DIP**	**UAV**	**IRT**	**PRS**
Technical Features	**Mapping stains**	Only suitable with other technologies (−)	Suitable (+)	Suitable (+)	Only suitable for some stains (−)	Suitable (+)
	Mapping cracking	Suitable (+)	Suitable (+)	Suitable (+)	Only suitable for some cracking (−)	Suitable (+)
	Mapping loss of adhesion	Suitable, in case of façade geometry change (−)	Unsuitable, if there is no image change (−)	Unsuitable, if there is no image change (−)	Suitable (+)	Unsuitable, if there is no image change (−)
	Measure the crack thickness	Suitable (+)	Suitable (+)	Only suitable combined with other technologies (−)	Unsuitable (−)	Suitable (+)
	Quantify defect area	Only suitable combined with other technologies (−)	Suitable (+)	Only suitable combined with other technologies (−)	Only suitable combined with other technologies (−)	Suitable (+)
	Survey in tall buildings	Unsuitable, with acquisition from the ground (−)	Unsuitable, with acquisition from the ground (−)	Suitable (+)	Unsuitable, with acquisition from the ground (−)	Unsuitable, with acquisition from the ground (−)
	Access to the facade surface	Unsuitable, with narrow streets around (−)	Only suitable if image acquisition is done by drone (−)	Suitable (+)	Unsuitable, with narrow streets around (−)	Unsuitable, with narrow streets around (−)
	Weather conditions	Independent (+)	Dependent (−)	Dependent (−)	Dependent (−)	Dependent (−)

* Unsuitable (−). Suitable (+).

With the purpose of mapping stains in facades, all technologies are useful, except 3D laser scanning (TLS) because it could not produce colour information useful for mapping the defects [55]. Infrared thermography (IRT) is advisable for mapping claddings with loss of adhesion, cracking, and stains with moisture or another defect due to thermal gradients [61]. The measurement of crack thickness could be supported by TLS or photogrammetry (PRS) combined with digital image processing (DIP) for width around 0.2 mm. However, a high resolution camera is needed to get a good result [20,56,73]. To quantify defect area in facades, PRS and DIP are advisable. With the first technique, it is possible to measure over the image. In DIP, the defect areas can be obtained through several image processing techniques, supported by a software, for the operation to convert the pixel in the distance [56].

There are some parameters associated with the nature of inspection that also influence the selection of the best technology. For surveying high buildings and facade surfaces with difficult accessibility, like narrow streets, drones (UAV) are the more advisable technology. As presented before, the weather conditions significantly influence the technology capacity to map and quantify defects in facades, like windy conditions, presence of obstacles (e.g., trees), or solar radiation in facades [49,70]. In this sense, TLS could be a useful technique in these conditions, with the limitations presented before.

As discussed, several technologies should be used to improve the calculation of the severity of the degradation index (S_w), according to the type of defects in the building facades and the nature of the inspection. This aim can be achieved through better mapping and measure of each type of defect.

7. Conclusions

Visual inspection remains a very important tool for the assessment of the physical and visual condition of buildings' facades. However, there is some uncertainty associated with this analysis, namely in the quantification of defect areas, as shown in Sections 3 and 4.

This study intends to present some recommendations to automate the inspection process to reduce the uncertainty related with the visual assessment, namely reducing the subjectivity related to the inspector in assessing, mapping, and measuring the defects in building facades. In this sense, the use of new technologies can help significantly in the assessment of the degradation condition of claddings and, consequently, increase the reliability of an in-service diagnosis.

From the emerging technologies analysed, each one has different advantages and limitations, according to the literature. In short, this study reveals that there is no suitable technology for all types of defects. To quantify the degradation in facades through mapping and measuring defects, the technologies that are revealed to be more advisable were photogrammetry and digital image processing. These techniques are the most suitable for mapping defects, measuring the crack thickness, and quantifying defect areas. However, for mapping defects related to loss of adhesion, infrared thermography is more advisable and can show a hidden defect.

When parameters related to the nature of inspection action are considered, like the weather condition or the incidence of solar radiation in the façade, 3D laser scanning could be the best technology to overcome these limitations. In tall buildings and with difficult access to the surface façade, drones are recommended. Taking into account the type of defects in the building facade and the nature of inspection, these technologies are capable of automating the visual inspection, producing a more reliable and accurate diagnosis about the degradation condition of facades. Further analyses are needed concerning the real application of these technologies, evaluating the accuracy of the detection of the anomalies observed, and the reliability of estimating the severity of the degradation index (S_w), based on the information collected through these techniques.

Author Contributions: Conceptualization, I.S.D. and I.F.-C. Methodology, I.S.D. and I.F.-C. Formal analysis, I.S.D., I.F.-C. and A.S. Investigation, I.S.D. Writing—original draft preparation, I.S.D., I.F.-C. and A.S. Writing—review and editing, I.S.D., I.F.-C. and A.S. All authors have read and agreed to the published version of the manuscript.

Funding: This research was funded by the Portuguese Foundation for Science and Technology (FCT) through project BestMaintenance-LowerRisks (PTDC/ECI-CON/29286/2017).

Data Availability Statement: Data is contained within the article.

Acknowledgments: The authors gratefully acknowledge the support of CERIS Research Centre (Instituto Superior Técnico—University of Lisbon).

Conflicts of Interest: The authors declare no conflict of interest.

References

1. Ferraz, G.T.; de Brito, J.; de Freitas, V.P.; Silvestre, J.D. State-of-the-art review of building inspection systems. *J. Perform. Constr. Facil.* **2016**, *30*, 04016018. [CrossRef]
2. Pereira, C.; de Brito, J.; Silvestre, J.D. Harmonising the classification of diagnosis methods within a global building inspection system: Proposed methodology and analysis of fieldwork data. *Eng. Fail. Anal.* **2020**, *115*, 104627. [CrossRef]
3. Pereira, C.; de Brito, J.; Silvestre, J.D.; Flores-Colen, I. Atlas of defects within a global building inspection system. *Appl. Sci.* **2020**, *10*, 5879. [CrossRef]
4. Pereira, C.; Silva, A.; de Brito, J.; Ferreira, C.; Flores-Colen, I.; Silvestre, J. *Uncertainty and Risk Analysis in Inspection and Diagnosis, CIB W086 State of the Art Report*; University of Lisbon: Lisbon, Portugal, 2020.
5. Gasparoli, P.; Jornet, A.; Panato, E. Factory-made dry rendering mortars: Characterization and artificial weathering. In Proceedings of the International Workshop on Management of Durability in the Building Process, Politecnico di Milano, Milan, Italy, 25 June 2003.
6. Kus, H.; Nygren, K.; Norberg, P. In-use performance assessment of rendered autoclaved aerated concrete walls by long-term moisture monitoring. *Build. Environ.* **2004**, *39*, 677–687. [CrossRef]
7. Daniotti, B.; Iacono, P. Evaluating the service life of external walls: A comparison between long-term and short-term exposure. In Proceedings of the X International Conference on Durability of Building Materials and Components, Lyon, France, 17–20 April 2005; p. 8.
8. Wyatt, D. The contribution of FMEA and FTA to the performance review and auditing of service life design of constructed assets. In Proceedings of the 10th Durability of Building Materials and Components (10DBMC), Lyon, France, 17–20 April 2005; pp. 17–20.
9. Köliö, A.; Honkanen, M.; Lahdensivu, J. Corrosion propagation phase studies on finnish reinforced concrete facades. In *New Approaches to Building Pathology and Durability*; Delgado, J., Ed.; Springer: Singapore, 2016; pp. 75–98.
10. Serralheiro, M.I.; de Brito, J.; Silva, A. Methodology for service life prediction of architectural concrete facades. *Constr. Build. Mater.* **2017**, *133*, 261–274. [CrossRef]
11. Shohet, I.; Rosenfeld, Y.; Puterman, M.; Gilboa, E. Deterioration patterns for maintenance management—A methodological approach. In Proceedings of the 8th International Conference on Durability of Building Materials and Components (DBMC), Vancouver, BC, Canada, 30 May–3 June 1999; pp. 1666–16678.
12. Straub, A. Using a condition-dependent approach to maintenance to control costs and performances. *J. Facil. Manag.* **2002**, *1*, 380–395. [CrossRef]
13. Daniotti, B.; Lupica Spagnolo, S. Service life prediction for buildings' design to plan a sustainable building maintenance. In Proceedings of the SB07 Lisbon—Sustainable Construction, Materials and Practices: Challenge of the Industry for the New Millennium, Minho, Portugal, 12–14 September 2007; IOS Press: Lisbon, Portugal; pp. 515–521.
14. Bauer, E.; Kraus, E.; Silva, M.N.B.; Zanoni, V.A.G. Evaluation of damage of building facades in Brasília. In Proceedings of the XIII DBMC—International Conference on Durability of Materials and Components, Sao Paulo, Brazil, 2–5 September 2014; pp. 35–542.
15. Peng, K.C.; Feng, L.; Hsieh, Y.C.; Yang, T.H.; Hsiung, S.H.; Tsai, Y.D.; Kuo, C. Unmanned Aerial Vehicle for infrastructure inspection with image processing for quantification of measurement and formation of facade map. In Proceedings of the International Conference on Applied System Innovation (ICASI), Sapporo, Japan, 13–17 May 2017; Institute of Electrical and Electronics Engineers Inc.: Piscataway, NJ, USA; pp. 1969–1972.
16. Dorafshan, S.; Maguire, M. Bridge inspection: Human performance, unmanned aerial systems and automation. *J. Civ. Struct. Heal. Monit.* **2018**, *8*, 443–476. [CrossRef]
17. Serrat, C.; Banaszek, A.; Cellmer, A.; Gibert, V. Use of UAVs for technical inspection of buildings within the BRAIN Massive inspection platform. In *IOP Conference Series: Materials Science and Engineering*; Institute of Physics Publishing: Bristol, UK, 2019; Volume 471.
18. Tuutti, K. *Corrosion of Steel in Concrete*; Swedish Cement and Concrete Research Institute: Stockholm, Sweden, 1982.
19. Berberan, A.; Portela, E.A.; Boavida, J. Assisted visual inspection of dams for structural safety control. In Proceedings of the Hydro 2006: Maximizing the Benefits of Hydropower, Porto Carras, Greece, 25–27 September 2006.
20. Law, D.W.; Holden, L.; Silcock, D. The assessment of crack development in concrete using a terrestrial laser scanner (TLS). *Aust. J. Civ. Eng.* **2015**, *13*, 22–31. [CrossRef]
21. Serrat Piè, C.; Gibert Armengol, V.; Cellmer, A.; Banaszek, A. Quantitative comparison between visual and UAV-based inspections for the assessment of the technical condition of building facades. In *Research and Modelling in Civil Engineering 2018*; Katzer, J., Cichocki, K., Domski, J., Eds.; Koszalin University of Technology Publishing House: Koszalin, Poland, 2018; pp. 19–29.
22. Masiero, A.; Costantino, D. TLS for detecting small damages on a building façade. *ISPRS Ann. Photogramm. Remote Sens. Spat. Inf. Sci.* **2019**, *XLll–2/W11*, 831–836. [CrossRef]
23. Marković, M.; Laban, M.; Kuzmić, T.; Vujinović, M.; Draganić, S. Application of modern technologies in assessing facade condition of building structures. In Proceedings of the FIG Working Week 2020, Smart Surveyors for Land and Water Management, Amsterdam, The Netherlands, 10–14 May 2020.
24. Agnisarman, S.; Lopes, S.; Chalil Madathil, K.; Piratla, K.; Gramopadhye, A. A survey of automation-enabled human-in-the-loop systems for infrastructure visual inspection. *Autom. Constr.* **2019**, *97*, 52–76. [CrossRef]

25. Madureira, S.; Flores-Colen, I.; de Brito, J.; Pereira, C. Maintenance planning of facades in current buildings. *Constr. Build. Mater.* **2017**, *147*, 790–802. [CrossRef]

26. Bortolini, R.; Forcada, N. Building Inspection System for Evaluating the Technical Performance of Existing Buildings. *J. Perform. Constr. Facil.* **2018**, *32*, 04018073. [CrossRef]

27. Moropoulou, A.; Labropoulos, K.C.; Delegou, E.T.; Karoglou, M.; Bakolas, A. Non-destructive techniques as a tool for the protection of built cultural heritage. *Constr. Build. Mater.* **2013**, *48*, 1222–1239. [CrossRef]

28. McGibbon, S.; Abdel-Wahab, M. Emerging digitisation trends in stonemasonry practice. In Proceedings of the Science and Art: A Future for Stone: Proceedings of the 13th International Congress on the Deterioration and Conservation of Stone, Paisley, Glasgow, Scotland, 6–10 September 2016; Volume 2, pp. 1041–1050.

29. Calantropio, A.; Chiabrando, F.; Rinaudo, F.; Teppati Losè, L. Use and evaluation of a short range small quadcopter and a portable imaging laser for built heritage 3D documentation. *Int. Arch. Photogramm. Remote Sens. Spat. Inf. Sci. ISPRS Arch.* **2018**, *42*, 71–78. [CrossRef]

30. Markiewicz, J.; Tobiasz, A.; Kot, P.; Muradov, M.; Shaw, A.; Al-Shammaa, A. Review of surveying devices for structural health monitoring of cultural heritage buildings. In Proceedings of the International Conference on Developments in eSystems Engineering, DeSE, Kazan, Russia, 20 October 2019; Institute of Electrical and Electronics Engineers Inc.: Piscataway, NJ, USA; pp. 597–601.

31. Shariq, M.H.; Hughes, B.R. Revolutionising building inspection techniques to meet large-scale energy demands: A review of the state-of-the-art. *Renew. Sustain. Energy Rev.* **2020**, *130*. [CrossRef]

32. Valero, E.; Forster, A.; Bosché, F.; Hyslop, E.; Wilson, L.; Turmel, A. Automated defect detection and classification in ashlar masonry walls using machine learning. *Autom. Constr.* **2019**, *106*, 102846. [CrossRef]

33. Lee, D.Y.; Chi, H.L.; Wang, J.; Wang, X.; Park, C.S. A linked data system framework for sharing construction defect information using ontologies and BIM environments. *Autom. Constr.* **2016**, *68*, 102–113. [CrossRef]

34. De Brito, J.; Pereira, C.; Silvestre, J.; Flores-Colen, I. *Expert Knowledge-Based Inspection Systems: Inspection, Diagnosis, and Repair of the Building Envelope*; Springer International Publishing: New York, NY, USA, 2020; ISBN 978-3-030-42446-6.

35. Gonçalves, J.; Mateus, R.; Silvestre, J.D. Comparative analysis of inspection and diagnosis tools for ancient buildings. In Proceedings of the Digital Heritage, Progress in Cultural Heritage: Documentation, Preservation, and Protection. 7th International Conference, EuroMed 2018, Nicosia, Cyprus, 29 October–3 November 2018; Springer: Berlin/Heidelberg, Germany, 2018; Volume 11197 LNCS, pp. 289–298.

36. Van den Beukel, A. *Building Pathology: A State-of-the-Art Report*; CIB W086-Publication 155; Conseil International du Bâtiment: Rotterdam, The Netherlands, 1993.

37. De Freitas, V.P. *A State-of-the-Art Report on Building Pathology*; CIB W086-Publication 393; Conseil International du Bâtiment: Rotterdam, The Netherlands, 2013.

38. Pereira, C.; de Brito, J.; Silvestre, J.D. Harmonised classification of the causes of defects in a global inspection system: Proposed methodology and analysis of fieldwork data. *Sustainability* **2020**, *12*, 5564. [CrossRef]

39. Silvestre, J.D.; de Brito, J. Inspection and repair of ceramic tiling within a building management system. *J. Mater. Civ. Eng.* **2010**, *22*, 39–48. [CrossRef]

40. Neto, N.; de Brito, J. Inspection and defect diagnosis system for natural stone cladding. *J. Mater. Civ. Eng.* **2011**, *23*, 1433–1443. [CrossRef]

41. Sá, G.; Sá, J.; De Brito, J.; Amaro, B. Statistical survey on inspection, diagnosis and repair of wall renderings. *J. Civ. Eng. Manag.* **2015**, *21*, 623–636. [CrossRef]

42. Silva, A.; de Brito, J.; Gaspar, P.L. Methodologies for service life prediction of buildings: With a focus on façade claddings. In *Green Energy and Technology*, 1st ed.; Springer International Publishing: New York, NY, USA, 2016; ISBN 978-3-319-33288-8.

43. Silva, A.; de Brito, J. Do we need a buildings' inspection, diagnosis and service life prediction software? *J. Build. Eng.* **2019**, *22*, 335–348. [CrossRef]

44. Pereira, C.; de Brito, J.; Silvestre, J.D. Global inspection, diagnosis and repair system for buildings: Managing the level of detail of the defects classification. In Proceedings of the 7th Rehabend Congress—Construction Pathology, Rehabilitation Technology and Heritage Management, Cáceres, Spain, 24–27 March 2018; pp. 572–579.

45. Gaspar, P.L.; Brito, J. de Quantifying environmental effects on cement-rendered facades: A comparison between different degradation indicators. *Build. Environ.* **2008**, *43*, 1818–1828. [CrossRef]

46. Gaspar, P.L. Service Life of Constructions: Development of a Method to Estimate the Durability of Construction Elements. Application to Renderings of Current Buildings. Ph.D. Thesis, Instituto Superior Técnico, University of Lisbon, Lisbon, Portugal, 2009.

47. Ferreira, C.; Canhoto Neves, L.; Silva, A.; de Brito, J. Stochastic Petri net-based modelling of the durability of renderings. *Autom. Constr.* **2018**, *87*, 96–105. [CrossRef]

48. Silva, A.; De Brito, J.; Gaspar, P.L. Service life prediction model applied to natural stone wall claddings (directly adhered to the substrate). *Constr. Build. Mater.* **2011**, *25*, 3674–3684. [CrossRef]

49. Falorca, J.; Lanzinha, J.C.G. *Developments Towards the Use of Drones in the Building Envelope Condition Assessment: A Comprehensive Review and Experimental Rehearsals*, 1st ed.; Universidade da Beira Interior (UBI): Covilhã, Portugal, 2019; ISBN 978-989-654-610-6.

50. El Masri, Y.; Rakha, T. A scoping review of non-destructive testing (NDT) techniques in building performance diagnostic inspections. *Constr. Build. Mater.* **2020**, *265*, 120542. [CrossRef]

51. Smits, J. Application of 3D terrestrial laser scanning to map building surfaces. *J. Arch. Conserv.* **2011**, *17*, 81–94. [CrossRef]
52. Sánchez-Aparicio, L.J.; Del Pozo, S.; Ramos, L.F.; Arce, A.; Fernandes, F.M. Heritage site preservation with combined radiometric and geometric analysis of TLS data. *Autom. Constr.* **2018**, *85*, 24–39. [CrossRef]
53. Nowak, R.; Orłowicz, R.; Rutkowski, R. Use of TLS (LiDAR) for building diagnostics with the example of a historic building in Karlino. *Buildings* **2020**, *10*, 24. [CrossRef]
54. Suchocki, C.; Katzer, J. TLS Technology in brick walls inspection. In Proceedings of the 2018 Baltic Geodetic Congress (BGC Geomatics), Olsztyn, Poland, 21–23 June 2018; Institute of Electrical and Electronics Engineers Inc.: New York, NY, USA, 2018; pp. 359–363.
55. Russo, M.; Carnevali, L.; Russo, V.; Savastano, D.; Taddia, Y. Modeling and deterioration mapping of façades in historical urban context by close-range ultra-lightweight UAVs photogrammetry. *Int. J. Arch. Herit.* **2019**, *13*, 549–568. [CrossRef]
56. Valença, J.; Dias-Da-Costa, D.; Júlio, E.; Araújo, H.; Costa, H. Automatic crack monitoring using photogrammetry and image processing. *Meas. J. Int. Meas. Confed.* **2013**, *46*, 433–441. [CrossRef]
57. Kamnik, R.; Nekrep Perc, M.; Topolšek, D. Using the scanners and drone for comparison of point cloud accuracy at traffic accident analysis. *Accid. Anal. Prev.* **2020**, *135*, 105391. [CrossRef] [PubMed]
58. Edis, E.; Flores-Colen, I.; de Brito, J. Passive Thermographic inspection of adhered ceramic claddings: Limitation and conditioning factors. *J. Perform. Constr. Facil.* **2013**, *27*, 737–747. [CrossRef]
59. Kylili, A.; Fokaides, P.A.; Christou, P.; Kalogirou, S.A. Infrared thermography (IRT) applications for building diagnostics: A review. *Appl. Energy* **2014**, *134*, 531–549. [CrossRef]
60. Edis, E.; Flores-Colen, I.; De Brito, J. Passive thermographic detection of moisture problems in façades with adhered ceramic cladding. *Constr. Build. Mater.* **2014**, *51*, 187–197. [CrossRef]
61. Edis, E.; Flores-Colen, I.; De Brito, J. Building thermography: Detection of delamination of adhered ceramic claddings using the passive approach. *J. Nondestruct. Eval.* **2015**, *34*, 1–13. [CrossRef]
62. Bauer, E.; de Freitas, V.P.; Mustelier, N.; Barreira, E.; de Freitas, S.S. Infrared thermography–Evaluation of the results reproducibility. *Struct. Surv.* **2015**, *33*, 20–35. [CrossRef]
63. Bauer, E.; Pavón, E.; Pereira, C.H.F.; Nascimento, M.L.M. Criteria for identification of ceramic detachments in building facades with infrared thermography. In *Recent Developments in Building Diagnosis Techniques*; Delgado, J.M.P.Q., Ed.; Springer: Singapore, 2016; pp. 51–68.
64. Bauer, E.; Milhomem, P.M.; Aidar, L.A.G. Evaluating the damage degree of cracking in facades using infrared thermography. *J. Civ. Struct. Heal. Monit.* **2018**, *8*, 517–528. [CrossRef]
65. Lucchi, E. Applications of the infrared thermography in the energy audit of buildings: A review. *Renew. Sustain. Energy Rev.* **2018**, *82*, 3077–3090. [CrossRef]
66. Costanzo, A.; Minasi, M.; Casula, G.; Musacchio, M.; Buongiorno, M. Combined use of terrestrial laser scanning and IR thermography applied to a historical building. *Sensors* **2014**, *15*, 194–213. [CrossRef] [PubMed]
67. Lagüela, S.; Armesto, J.; Arias, P.; Herráez, J. Automation of thermographic 3D modelling through image fusion and image matching techniques. *Autom. Constr.* **2012**, *27*, 24–31. [CrossRef]
68. González-Jorge, H.; Lagüela, S.; Krelling, P.; Armesto, J.; Martínez-Sánchez, J. Single image rectification of thermal images for geometric studies in façade inspections. *Infrared Phys. Technol.* **2012**, *55*, 421–426. [CrossRef]
69. FLIR Products. Available online: https://www.flir.eu/products/tg267/ (accessed on 21 December 2020).
70. Rakha, T.; Gorodetsky, A. Review of Unmanned Aerial System (UAS) applications in the built environment: Towards automated building inspection procedures using drones. *Autom. Constr.* **2018**, *93*, 252–264. [CrossRef]
71. Garrido, I.; Lagüela, S.; Otero, R.; Arias, P. Thermographic methodologies used in infrastructure inspection: A review—Post-processing procedures. *Appl. Energy* **2020**, *266*, 114857. [CrossRef]
72. Valença, J.; Júlio, E.N.B.S.; Araújo, H.J. Applications of photogrammetry to structural assessment. *Exp. Tech.* **2012**, *36*, 71–81. [CrossRef]
73. Jiang, R.; Jáuregui, D.V.; White, K.R. Close-range photogrammetry applications in bridge measurement: Literature review. *Measurement* **2008**, *41*, 823–834. [CrossRef]
74. Oliveira, A.; Boavida, J.; Santos, B. Displacement and surface pathology monitoring of former Tejo Power Station building by combining terrestrial laser scanning, micro-geodesy, photogrammetry and GIS. In Proceedings of the INGEO 2017—7th International Conference on Engineering Surveying, Lisbon, Portugal, 18–20 October 2017; pp. 1–8.
75. Berberan, A.; Portela, E.A.; Boavida, J. Enhancing on-site dams visual inspections. In Proceedings of the 5th International Conference on Dam Engineering, Lisbon, Portugal, 14–16 February 2007; pp. 55–62.
76. Paulo, P.V.; Branco, F.A.; de Brito, J. Using orthophotography based on buildingslife software to inspect building facades. *J. Perform. Constr. Facil.* **2014**, *28*, 04014019. [CrossRef]
77. Agisoft Metashape Software. Available online: https://www.agisoft.com/buy/online-store/ (accessed on 21 December 2020).
78. Laofor, C.; Peansupap, V. Defect detection and quantification system to support subjective visual quality inspection via a digital image processing: A tiling work case study. *Autom. Constr.* **2012**, *24*, 160–174. [CrossRef]
79. González Manich, C.; Kelman, T.; Coutts, F.; Qiu, B.; Murray, P.; González-Longo, C.; Marshall, S. Exploring the use of image processing to survey and quantitatively assess historic buildings. In *Structural Analysis of Historical Constructions: Anamnesis, Diagnosis, Therapy, Controls*; Koen Van Balen, E.V., Ed.; CRC Press: Leuven, Belgium, 2016; pp. 125–132. ISBN 9781138029514.

80. Ruiz, R.D.B.; Lordsleem Júnior, A.C.; Fernandes, B.J.T.; Oliveira, S.C. Unmanned aerial vehicles and digital image processing with deep learning for the detection of pathological manifestations on facades. In Proceedings of the 18th International Conference on Computing in Civil and Building Engineering, ICCCBE 2020, Lecture Notes in Civil Engineering, São Paulo, Brazil, 18–20 August 2020; Santos, E.T., Scheer, S., Eds.; Springer: Berlin/Heidelberg, Germany, 2021; Volume 98, pp. 1099–1112.
81. Pereira, F.C.; Pereira, C.E. Embedded image processing systems for automatic recognition of cracks using UAVs. *IFAC Pap.* **2015**, *28*, 16–21. [CrossRef]
82. DJI Products. Available online: https://store.dji.com/pt/product/phantom-4-pro-v2?from=landing_page&site=brandsite&vid=43151 (accessed on 21 December 2020).
83. Mosly, I. Applications and issues of unmanned aerial systems in the construction industry. *Int. J. Constr. Eng. Manag.* **2017**, *6*, 235–239. [CrossRef]
84. Eschmann, C.; Kuo, C.M.; Kuo, C.H.; Boller, C. Unmanned aircraft systems for remote building inspection and monitoring. In Proceedings of the 6th European Workshop on Structural Health Monitoring (EWSHM 2012), Dresden, Germany, 3–6 July 2012; Volume 2, pp. 1179–1186.
85. Chen, K.; Reichard, G.; Xu, X. Opportunities for applying camera-equipped drones towards performance inspections of building facades. In Proceedings of the Computing in Civil Engineering 2019: Smart Cities, Sustainability, and Resilience—Selected Papers from the ASCE International Conference on Computing in Civil Engineering 2019, Atlanta, GA, USA, 17–19 June 2019; American Society of Civil Engineers (ASCE): Reston, VA, USA; pp. 113–120.
86. Furtado, F.J.; Gonçalves, L.J.C. Facade inspections with drones—Theoretical analysis and exploratory tests. *Int. J. Build. Pathol. Adapt.* **2020**, *24*. [CrossRef]
87. Roca, D.; Lagüela, S.; Díaz-Vilariño, L.; Armesto, J.; Arias, P. Low-cost aerial unit for outdoor inspection of building façades. *Autom. Constr.* **2013**, *36*, 128–135. [CrossRef]

 *buildings*

Review

Microclimate of Air Cavities in Ventilated Roof and Façade Systems in Nordic Climates

Sara Bredal Ingebretsen, Erlend Andenæs * and Tore Kvande

Department of Civil and Environmental Engineering, Norwegian University of Science and Technology, 7491 Trondheim, Norway; sarabin@stud.ntnu.no (S.B.I.); tore.kvande@ntnu.no (T.K.)
* Correspondence: erlend.andenas@ntnu.no

Abstract: Accurate values for the climatic conditions in an air cavity, hereby called the microclimate, are crucial when calculating and simulating the performance of a ventilated roof and façade system. The climatic stress of its components and their mould and rot potential influence the long-term durability of the roof or façade. A scoping study is conducted to gain an overview on research and the scientific literature on the microclimate of air cavities in ventilated roofing and claddings in Nordic climates. From the body of the research literature, 21 scientific works were of particular interest, and their findings are summarized. The review shows that only a limited number of studies discuss the microclimate of air cavities. Roofs are discussed to a greater and more varied degree compared to façades and air cavities behind solar panels. However, the results cannot be compared and validated against each other to generally describe the microclimate of air cavities, as the surveyed papers approach the subject differently. This knowledge gap indicates that calculations and simulations can be performed without knowing whether the results represent reality. If the structure of ventilated roof and façade systems are only designed based on experience, it can be difficult to be proactive and adapt to future climate changes. Further studies are needed to determine the relation between the exterior climate and the air cavity microclimate, so that future climate predictions can be used to simulate the long-term performance of ventilated roof and façade systems.

Keywords: air cavity; air gap; ventilated roof; ventilated façade system; Nordic climate; literature survey

Citation: Ingebretsen, S.B.; Andenæs, E.; Kvande, T. Microclimate of Air Cavities in Ventilated Roof and Façade Systems in Nordic Climates. *Buildings* **2022**, *12*, 683. https://doi.org/10.3390/buildings12050683

Academic Editor: Ana Silva

Received: 23 March 2022
Accepted: 16 May 2022
Published: 19 May 2022

Publisher's Note: MDPI stays neutral with regard to jurisdictional claims in published maps and institutional affiliations.

1. Introduction

The climate screen of a building mainly consists of its outer walls and roof. Its primary function is to create a shelter against weather exposure. The type and severity of this exposure depends on the geographical location of the building and local conditions specific to its site. In a Nordic climate, the weather will involve strong winds, precipitation, snow loads, great temperature fluctuations, and freeze–thaw cycles [1]. The climate is expected to become more severe in coming years. The temperature and annual precipitation will increase, and extreme precipitation events will become more intense and more common [2]. Climate change of this type will affect material durability, due to an increase in conditions where wood materials will be at risk of deterioration due to biological growth [3]. Among other reasons, the reduction in the number and duration of frost periods is particularly unfortunate, as the growth of mould and fungi will be drastically impeded during and after a frost period [4,5]. The climate zones of the Nordic countries are illustrated in Figure 1.

Figure 1. Climate Classification map for Nordic countries according to the Köppen–Geiger system—all cities above 100,000 inhabitants are marked (2016). Reprinted with permission from Ref. [6].

Specific climatic challenges in the Nordic climate include wind-driven rain, freeze–thaw cycles, strong winds, frost, low annual average temperatures, and snow loads, often in frequently varying combinations. Especially in coastal areas, the weather can change rapidly and vary greatly over short periods. To adapt to the varied Nordic climate, facades and sloped roofs are usually built according to a principle of two-stage weatherproofing ("totrinnstetning" in Norwegian) [7]. A universally accepted English translation of this term has not been identified, but the authors are aware that the terms "two-step" or "two-layer" weatherproofing have also been in use, as well as "two-stage tightening" or "two-stage sealing/seals". The term "two-stage weatherproofing" is preferred in this article as multiple material layers may be involved in the assembly, and the terms "tightening" or "sealing" are direct, but erroneous, translations of the Norwegian term. A roof or façade built according to this principle consists of a rain screen (cladding) and a vapour-open wind barrier layer, separated by an air cavity for drainage and ventilation. The air cavity ensures that precipitation does not leak into the building envelope, and that any moisture in the building envelope is allowed to dry. Note that the air cavity is ventilated to the exterior climate to remove moisture, as opposed to systems such as Trombe walls that are ventilated to the building's interior for purposes of indoor ventilation [8,9]. For roofs, the ventilated air cavity also prevents heat flows from the building interior from melting snow piled up on the roof tiles, reducing the risk of ice formation. The principle of two-stage weatherproofing is illustrated in Figure 2.

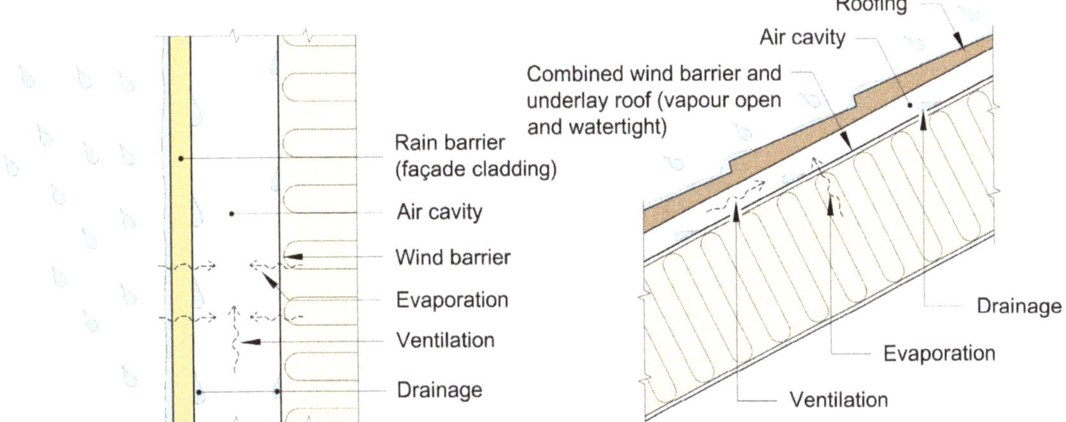

Figure 2. Illustration of the principle of two-stage weatherproofing for a façade (**left**) and a roof (**right**). Reprinted with permission from Refs. [10,11].

The principle of two-stage weatherproofing has been shown to create robust roofs and façades [7], but climate change might alter its performance. Norwegian statistics show that the share of process-induced building defects caused by precipitation have increased from 24% for the period 1993–2002 to 42% for the period 2017–2020 [12]. Insufficient ventilation and/or drainage of the air cavity is often the main reason why precipitation causes defects in a ventilated façade. Although flaws of the ventilation cavity are usually caused by errors in design or execution that deviate from good moisture safety practices, defects have also been reported in façades built in a way traditionally considered sound [13]. More humid weather with more precipitation and shorter frost periods causes reports of moisture damages hitherto unseen. Wooden battens rot while the cladding remains intact. In a new climate, existing knowledge of moisture-safe design seems to not apply anymore.

Being able to predict service life and making sound material choices for the building envelope is a central part of the building physical design of a building. Service life prediction is conducted using accelerated artificial ageing experiments [14], numerical simulations [15], and knowledge gathering [16,17]. Traditionally, assessments of durability are conducted using the exterior climate as an input parameter. Assessments of accelerated artificial ageing methods show that the determination of precise boundary conditions is a challenge even for façades with only single-stage weatherproofing [18]. Assessing the durability of materials exposed to the microclimate of the air cavity will likewise require in-depth understanding of the climatic boundary conditions. The microclimate of the air cavity will not necessarily correlate directly with the exterior climate [19]. The term "microclimate" is defined as a set of climatic conditions in a limited area, which may differ from those of the surrounding area [20]. New experiences and observed defects suggest a need for better knowledge of climatic conditions in the air cavity to accurately assess the durability of battens, wind barriers, and wind barrier tape. Relevant microclimate parameters for the air cavity may include temperature, relative humidity, air pressure, water intrusion, and air flow characteristics, and how these parameters all behave in relation to those of the exterior climate. This knowledge will be useful when selecting the methodology for accelerated artificial ageing and for numerical simulations of mould/rot risk throughout the lifetime of buildings in Nordic climates.

The purpose of this study is to map the current body of knowledge regarding the long-term microclimate of the air cavity in a roof or wall façade following the two-stage principle for weatherproofing. To address this general issue, the following research questions are investigated:

1. To what degree are the climatic conditions of the air cavity of two-stage weather-proofed façades addressed in the research literature?
2. Which climatic conditions can be expected in the air cavity of a two-stage weather-proofed façade?
3. What are the most important knowledge gaps concerning the climatic conditions of air cavities?

To answer these research questions, a systematic review of the research literature was conducted, using different library databases. Note that there appears to be some disagreement on the nomenclature of the term "air cavity", with many sources using the term "air gap" instead. This article consistently uses the term "air cavity", even when reporting on literature using "air gap" in the original text, provided that the cavity in question is ventilated at the top and bottom (an "inlet" and "outlet" opening).

Some limitations to the study are acknowledged: A considerable portion of the literature originates from Norway. This is not unexpected, as the principle of two-stage weatherproofing originates from the Norwegian Building Research Institute (later incorporated into SINTEF) [21–23], where research and development of the principle continues today. Other northern European countries, as well as Canada and the United States, have also adopted the construction principle to various degrees. However, in Canada, the main function of the air cavity appears to be drainage, and air cavities are built considerably narrower than in Nordic countries—as narrow as 1 mm [24,25]. For this reason, research from North America was excluded from the main part of this study. Research on air cavities was also identified from Central Europe, but this was also excluded due to the climate being different from Nordic countries. The warmer Central European climate leads to different considerations to be taken when assessing durability and service life. Nevertheless, relevant studies from countries beyond the Nordics are discussed in brief at the end of the Results chapter.

2. Methodology

2.1. Scoping Studies

To investigate the current body of knowledge on the microclimate of air cavities, a scoping study was conducted. Scoping studies are used to rapidly map the key terms and concepts within a specific area of research. The framework developed by Arksey and O'Malley [26] was used to guide the literature search through the following five stages:

1. Identifying the research questions;
2. Identifying relevant studies;
3. Study selection;
4. Charting the data;
5. Collating, summarizing, and reporting the results.

The specific procedure used in this study is a refinement of the framework described in [27]. The full procedure is illustrated in Figure 3.

2.2. Literature Search Parameters

The search was conducted through the following literature databases: Scopus, ScienceDirect, Google Scholar, and Web of Science. For an overview of the search terms and search strings that were employed, see the Supplementary File to this article. Filters in the search databases were used to limit the results to scientific articles in English. To further ensure that results were relevant to Nordic climates, the selection of articles was limited to articles originating from Denmark, Finland, Iceland, Ireland, Norway, Sweden, and the United Kingdom. This selection was made to limit the relevance to a Nordic climate, with its associated climatic parameters such as the frequent combination of precipitation and wind, freeze–thaw cycles, frost, and low annual average temperatures, as opposed to cold climates in general which may include high-altitude locations in warmer countries or entirely polar climates.

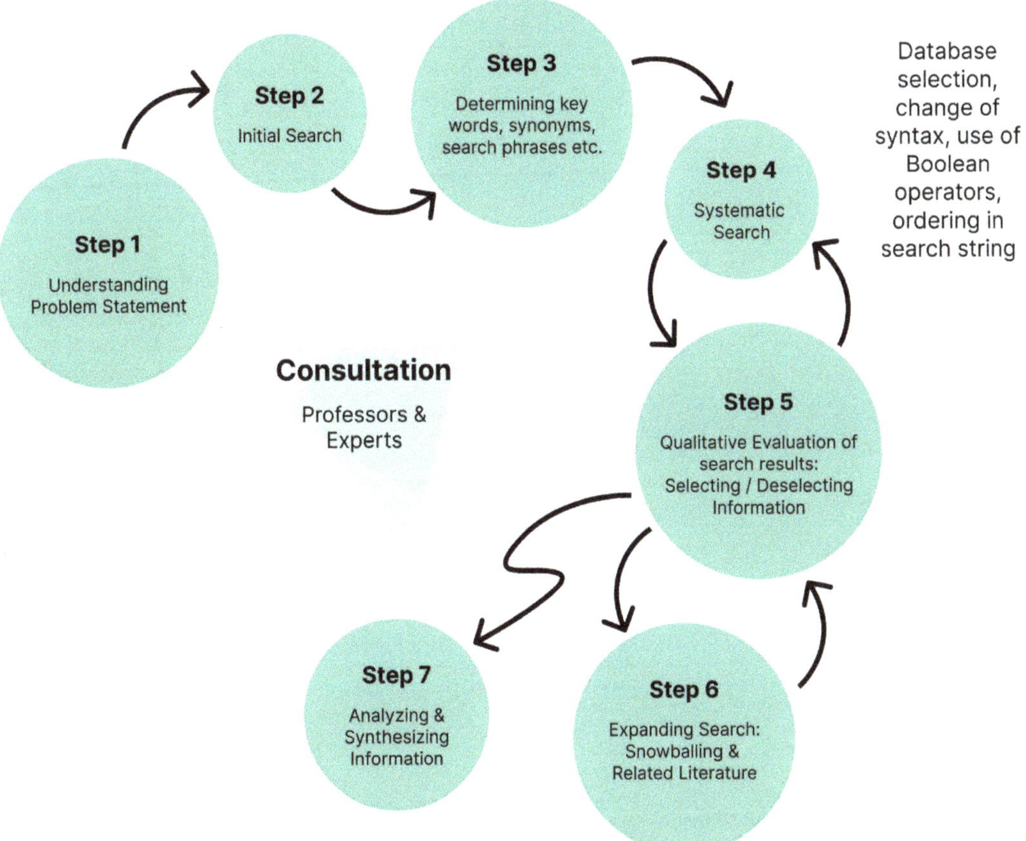

Figure 3. Scoping study methodology. Reprinted with permission from Ref. [28].

After a list of relevant articles was found through the database search, citation chaining, or "snowballing" [29], was used to find further relevant studies that were not found in the database search. Search results (Step 5 in Figure 3) were reviewed by assessing the titles of each study. For articles that seemed relevant, abstracts were thoroughly read, and if they appeared to match the research questions, they were included for further assessment of relevance. The search procedure thus resulted in a list of 70 relevant articles, of which 54 were discarded from the final study, for one or more of the following reasons:

- The article did not discuss the microclimate of air cavities (e.g., [30]).
- A microclimate was discussed, but not in relation to a ventilated façade (e.g., [31]).
- The research design of the article did not contain an air cavity, instead focusing on the properties of materials such as insulation or the cladding itself (e.g., [32]).
- The article featured temperature and moisture measurements, but only from the outside of the cladding or on the inside of the wind barrier (e.g., [33,34]).
- The article discussed air cavities that were not open to the exterior air at both the top and bottom of the façade (e.g., [8,9]).
- The article only considered masonry structures (e.g., [35]).
- Review articles of studies that chiefly contained studies from other climates or countries (e.g., [36]).

- Articles based on results from irrelevant climates (humid tropical climates, desert climates, etc.) (e.g., [8,9,37]).

Since articles from Norwegian universities constituted a large part of the relevant literature, an additional search was conducted in the Norwegian library database to find M.Sc. theses or other reports that could possibly contain relevant information. This additional search identified five theses containing relevant works of research, primarily published in Norwegian.

3. Literature Review Results

3.1. General Overview of the Literature

The literature search identified 16 scientific articles and 5 M.Sc. theses that discuss the microclimate of ventilated façade and roof cavities specifically in a Nordic climate context. The review includes articles focusing on field studies, laboratory studies, computer simulations, and/or calculations. Of the 21 works of research, 6 focus on façades, 10 focus on sloped roofs, 3 concern roofs and/or façades with integrated photovoltaics, and 2 focus on both façades and roofs. A further 11 studies from outside the Nordic climate were selected to provide a brief overview of relevant research internationally, and to take in some perspectives that were not discussed by the Nordic articles. However, the limits to their applicability in a Nordic context must be noted.

The relevant research originated from Norway (Norwegian University of Science and Technology and SINTEF (formerly Selskapet for Industriell og Teknisk Forskning) in Trondheim), Sweden (Gävle University College, Kungliga Tekniska Högskulan in Stockholm, Lund University, and Chalmers University of Technology in Gothenburg), and Finland (Aalto University near Helsinki). Relevant studies were sought, but not found, from Denmark, Iceland, Ireland, and the United Kingdom, whose climates are similar.

Table 1 presents an overview of the individual studies identified in the literature review. Some studies resulted in (or contributed to) multiple publications, as shown in the "References" column. The investigated climate parameters of each study are also listed, as well as the duration of measurements (if applicable). Note that not all of the studies measured the parameters the same way or presented the measurements directly.

Table 1. Overview of studies.

Type of Study	References	Exterior Climate	Duration	Main Purpose of Study	Analysed Parameters of Relevance
Field study	[38–40]	Lund, Sweden	4 months	Describing cavity air flows and drying potential. Validating numerical model.	- Air velocity (measured); - Temperature (measured); - Exterior temperature (measured); - Exterior wind speed and approach angle (measured); - Driving forces (calculated); - ACH (calculated); - Drying rates (calculated).
Field study	[41]	Trondheim, Norway	-	Describing the impact of wind on heat transmission through wooden walls.	- Wind pressure (measured); - Wind pressure coefficients and gradient (calculated); - Exterior wind speed and approach angle (measured).

Table 1. *Cont.*

Type of Study	References	Exterior Climate	Duration	Main Purpose of Study	Analysed Parameters of Relevance
Field study	[42]	Trondheim, Norway	4 years	Describing factors affecting the hygrothermal performance of wooden claddings in Norway.	- Exterior temperature (measured); - RH in exterior air (measured); - Exterior wind speed and approach angle (measured); - Global radiation (measured); - Wind-driven rain (measured); - Surface temperature in air cavity (measured); - RH (measured); - Water vapour pressure (calculated); - Moisture content in wood (measured).
Simulations (validating field study)	[43]			Studying wind-induced airflows in cavities.	- Air flow characteristics (calculated); - Air cavity speed (calculated); - ACH (calculated).
Field study	[44]	Espoo, Finland	2.5 years	Evaluating a façade with exterior insulation.	- Temperature (measured); - RH (measured); - Mould index (calculated).
Laboratory study	[45]	Interior lab climate	-	Describing air flows in cavities and assessing how design details affect pressure losses.	- Air flow characteristics (calculated/measured); - Air pressure (measured); - Temperature (measured).
	[46]				- Air flow characteristics (calculated/measured); - Air pressure (measured); - Temperature (measured).
	[47]				- Air flow characteristics (calculated/measured); - Temperature (measured); - Air pressure (measured).
	[48,49]	Interior lab climate with simulated solar radiation.	-	Describing air flow characteristics and temperature conditions in roof cavities.	- Temperature (measured); - Air velocity (measured); - Thermal buoyancy (calculated).
	[50]			Modelling air flows in roof cavities due to buoyancy.	- Temperature (measured); - Surface temperature in air cavity (measured); - Air flow characteristics (calculated/measured); - Thermal buoyancy (calculated).
Simulations (validating Laboratory study)	[51]	-	-	Numerical study of roof cavity air flows.	- Air flow characteristics (calculated); - Temperature (calculated).
Field study	[52]	Skellefteå, Stockholm, and Växjö, Sweden	1–2 years (multiple)	Evaluating a simulation model for mould growth in roof cavities.	- RH (measured); - Temperature (measured); - RH_{crit} (calculated).

Table 1. *Cont.*

Type of Study	References	Exterior Climate	Duration	Main Purpose of Study	Analysed Parameters of Relevance
Field study	[53]	Trondheim, Norway	8–12 days (multiple)	Evaluating condensation risk in roof cavity. Relating cavity air flows to wind.	- Exterior temperature (measured); - Exterior wind speed and approach angle (measured); - Surface temperatures in air cavity (measured); - Temperature (measured); - Air velocity (measured).
	[19]		6–11 days (multiple)	Examining the influence of temperature and wind on roof ventilation. Condensation risk.	- Exterior temperature (measured); - Exterior wind speed and approach angle (measured); - Surface temperatures in air cavity (measured); - Temperature (measured); - Air velocity (measured); - Condensation potential (calculated).
	[54]		9 days/21 days	Mapping the influence of exterior climate on air cavity climates.	- Temperature (measured); - RH (measured); - Wood moisture (measured); - Exterior temperature (measured); - RH in exterior air (measured); - Exterior wind speed and approach angle (measured); - Global radiation (measured).
Laboratory study	[55–57]	Interior lab climate with simulated solar radiation.	-	Analysing air flows and heat transfer behind solar panels on façades.	- Temperature (measured); - Surface temperature in air cavity (measured); - Surface temperature (measured); - Air flow rate (measured); - Air velocity (measured); - Aspect ratio of air cavity.

3.2. Façades

Falk and Sandin [38] described field studies measuring air velocities and temperatures in south-facing, vented, and ventilated façades using different batten configurations. The research period spanned 2400 h between the beginning of October and the middle of February. A single sensor was moved periodically between the three test specimens. The results were compared to a simulation model. The studies found an average absolute air velocity in the cavity of 0.195 m/s and an average driving force of 0.3–0.5 Pa. According to calculations, the air change rate for the cavity was 230-310 ACH (air change rate per hour). The maximum rate was 2–3 times higher than the average rate, primarily driven by thermal buoyancy. The results showed a clear correlation between average cavity temperature and flow characteristics, where the temperature increased if the flow resistance was increased.

A further study by Falk and Sandin [39] explored the connection between air cavity design, air change rate, and ventilation drying processes. Calculations and mathematical models were employed and compared to field studies from the previous article from Università di Napoli Federico II. The results showed that the model used to calculate ACH yielded values that agreed with field measurements. The calculations of drying rates showed that the evaporation of moisture to the air cavity varied greatly between the different air cavity designs, but the variations tended to even out over time. The measurements also showed that a favourable exterior climate and the colour of the façade cladding had a greater impact than the design of the air cavity, and that a cavity width of 40 mm yielded 20–25 times more drying potential than a cavity 5 mm wide.

A third study in this series [40] presented a method to predict ACH in an air cavity. Mathematical models were used alongside climate data and validated using field measurements from [38]. The results showed that the calculated ACH largely fit the measured values from the field studies, even reflecting hourly variations. The model still was shown to be slightly lacking for daytime conditions. Significant hourly deviations were observed during daytime hours, showing thermal buoyancy to be weak compared to wind pressures.

Uvsløkk [41] presented information on how wind affects the air pressure in an air cavity. Field experiments, laboratory experiments, and calculations were conducted. The results indicated that the geometry of the air cavity and the details near the cavity openings greatly affected the pressure gradients in the air cavity and that the calculation methods were less suitable for complex cavity geometries. The article concluded that for most locations in Norway, it will be sufficient to use a reference wind velocity of 10 m/s, 10 m above the ground, for calculations spanning the heating season.

Nore [42] sought to increase the scientific knowledge of the performance of wooden claddings through conducting field studies. Sensors were mounted on the inside and outside of the wooden cladding on a test building in Trondheim, measuring the exterior climate, the relative humidity in the air cavity, and the moisture content of the wooden cladding. Measurements ran across four years between 2004 and 2007. The measurements showed that the cladding's hygrothermal response depended greatly on the orientation of the façade, largely due to different exposure to solar radiation, wind, and wind-driven rain. Further, the results showed that an air cavity was necessary, but that an air cavity opening of 23 mm did not significantly improve the drying potential of a west-facing façade compared to a smaller cavity opening of 4 mm. The surface treatment of the cladding affected the response to climate exposure, where untreated wood responded faster to variations compared to surface-treated wood, giving larger fluctuations in the moisture content of the cladding. Changes in the claddings' moisture content were more correlated with variations in temperature, solar radiation, and wind speed than with variations in precipitation, wind-driven rain, and wind direction.

Nore et al. [43] studied the wind-driven air stream in a 23 mm air cavity in a ventilated cladding using RANS CFD simulation, a validation study, and a comparison to earlier studies. The validation study showed a margin of error of 25% for cavity air flows with a Reynold's number (Re) between 1000 and 3500. Simulations with a Re outside these parameters showed an ACH/U10 between 120 and 250 for a high reference wind speed of 10 m/s perpendicular to the façade. The CFD results of cavity air speed and air change rate were compared to those of previous experimental studies, indicating agreement. Local pressure losses were calculated by comparing the results of separate and combined simulations. The pressure loss coefficient for wind perpendicular to the façade was between 4.5 (middle cavities) and 5.7 (edge cavities), and for wind angled at 22.5 degrees, it was between 4.9 and 7.1.

Viljanen et al. [44] evaluated the performance of the air cavity for a highly insulated façade by assessing mould growth potential as a performance criterium and investigating which factors affect this performance. Field studies were conducted on two externally insulated air cavities with a thermal resistance $R = 0.18$ and $1.57\ m^2K/W$ on the exterior side. Measured parameters included absolute humidity, solar radiation, mould growth potential, and the temperature and relative humidity in the air cavity and the exterior air. The results indicated that the average temperature in the winter season was 0–2 °C lower in the air cavity than in the exterior air, but the RH was typically 0–15% lower, depending on the type of wall. The computational analyses showed that the optimal air change rate was 4–$40\ h^{-1}$. Significant mould growth in the air cavity was only possible near the inlet area, where the temperature was the lowest and RH the highest. The analytical model identified that the most influential factors for the hygrothermal conditions in the air cavity were the interior and exterior R values, the ACH of the air cavity, and the vapour resistance of the layers close to the air cavity. Insulation on the exterior side of the air cavity improved the performance of the air cavity, but it was shown that reducing the size of the inlet can be beneficial. Exterior R values should not exceed $R = 0.35$–$0.5\ m^2K/W$ to maintain solar-radiation-driven drying and to prevent mould growth during the summer season.

Overall, it may be seen that the research regarding the microclimate of façade air cavities primarily concerns air flows by buoyancy and responses to wind, while parameters such as moisture and temperature remain little studied. Viljanen's work [44] is a welcome exception, but unfortunately, the measurements were conducted on an unconventional wall assembly that cannot be directly compared to the remainder of the studies.

3.3. Roofs

Gullbrekken et al. [45] presented calculations to determine pressure losses at the inlet (eave) opening and inside the air cavity, consisting of friction losses and passing of tile battens. The study consisted of a large-scale laboratory experiment and a numerical analysis using the COMSOL software. The results showed that the pressure loss coefficient of the airstream was greatly dependent on the shape of the tile battens (aligned perpendicular to the air flow through the air cavity). Tile battens with rounded edges were found to incur a pressure loss coefficient 40% lower than battens with right-angled edges. Additionally, the results indicated that a classic eave design gave a lower pressure loss coefficient than modern eave design. Installing a bug net at the openings of the air cavity doubled the pressure loss coefficient. The calculations as well as laboratory and field studies were further described in an M.Sc. thesis by Hansen [46]. Hansen noted that the air cavity width should be increased by increasing the height of the longitudinal battens rather than the cross battens, to reduce pressure losses of the cavity air flow. Another M.Sc. thesis that contributed to the results of [44] is that of Eggen and Røer [47], who built upon the study by Hansen. The two theses disagree slightly on whether it is more advantageous to ventilate the roof cavity through openings in the eaves or behind the roof gutter. The disagreement might be explained by the different placement of sensors in the two experiments.

Bunkholt et al. [48] investigated, through a laboratory study, how the air flow through a ventilated roof cavity is affected by temperature and air cavity design. Measurements indicated that the air cavity temperature and flow conditions were affected by the cavity width as well as the roof angle. The temperature decreased and air flow rate increased with higher roof angles and larger cavity widths. Increasing the cavity width only increased the air temperature up to a certain point. The optimal cavity width was found to be 48 mm. Increasing the cavity width from 23 mm to 70 mm reduced thermal buoyancy by two thirds. The results of the study were also presented in a Master thesis [49]. A result reported in the thesis is that thermal buoyancy caused a driving pressure equivalent to a wind speed of 0.4–2 m/s. Thermal buoyancy may thus be the primary driving force of the airflow through the air cavity large portions of the year in Norway.

Säwén et al. [50] presented an analytical model to predict the air flow due to buoyancy in a ventilated roof assembly. The developed Thermal Buoyancy Model was compared to a laboratory study and a numerical calculation by using all three methods to analyse the air flow and thermal conditions in an air cavity using the same test setup and the same Norwegian climate. Results from the laboratory experiment and the numerical simulations quantitatively showed that an increase in the heat loads or the roof inclination increased the air flow rate due to increased driving forces. Increasing the cavity width increased the air volume flow due to decreased air flow resistance. The analytical model underestimated the air volume flow by an average of 20% compared to the laboratory experiment, and overestimated by 3% compared to the numerical calculations, but the trends are similar.

Mundt-Pedersen and Harderup [52] investigated the suitability of a 1D hygrothermal calculation tool in the design phase to prevent and evaluate the risk of mould growth and moisture damage in ventilated roofs. Moisture and temperature sensors were mounted in different positions in the roofs of four residential buildings in Sweden, while similar roof structures were modelled in the WUFI simulation program. The sensors measured temperature and RH over a period of 12–33 consecutive months. Quantitative results from sensors and WUFI were compared to each other and to critical values for mould growth. The results indicated that WUFI can be used as a tool to predict the climatic conditions, but that multiple parameters will affect the results and need to be considered to reliably assess the mould growth potential.

In addition to their evaluations of façades as described in Section 3.3., Viljanen et al. [44] assessed the performance of a roof structure. Field measurements were conducted similarly to what was carried out for façades, insulating the exterior side of a roof air cavity with a thermal resistance value R = 0.13 and 2.13 m^2K/W. Calculation analyses showed that the optimal air change rate for the setup was 20 h^{-1}. Significant mould growth in the cavity was only possible near the inlet, where the temperature was lowest and the RH was highest. The analytical model identified the most influential parameters for the hygrothermal conditions in the cavity to be the internal and external R values, the ACH of the cavity airspace, and the vapour permeability of the vapour control layer. Insulating the exterior side of the air cavity was shown to increase the performance of the air cavity. Exterior R values should not exceed 0.7 m^2K/W to maintain solar drying and to prevent mould growth in the summer season.

Gullbrekken et al. [53] described the risk of condensation on the interior side of the air cavity and mapped the relation between the air velocity in the cavity and the exterior wind speed. Field measurements were conducted in Norway using sensors to measure surface temperatures, the air temperature in the air cavity, and the air velocity in the cavity for three seasons: spring, summer, and autumn. The temperature below the roofing was shown to be lower than the exterior temperature 51%, 14%, and 56% of the time, respectively, for the three seasons. The measurements indicated a strong correlation between the wind speed and the air flow velocity in the air cavity. The hourly air exchange rate could be estimated as 11 h^{-1} for periods of low wind speed and 84 h^{-1} for periods of high wind speeds.

Bunkholt et al. [19] investigated the influence of exterior temperatures on the temperature in the air cavity, what condensation potential results from an air cavity temperature lower than the exterior temperature, and what the consequences of condensation risk can be. Field studies were conducted using sensors to measure surface temperatures, air temperature in the air cavity, and the air volume flow in the air cavity. Measurements were conducted over five periods in Norway between 2016 and 2018: one period each for spring, summer, and autumn, and two periods during winter. The results showed marked periods of air cavity temperatures being lower than the exterior air temperature and positive CP_i (condensation potential) values for extended periods, especially during spring and autumn, and in winter periods without snow on the roof. This indicated that the materials in the roof absorbed moisture and regulated the air humidity in the air cavity. Large negative CP_i values indicated drying of the roof structure.

Rønningen [54] conducted a field study on air cavities on research buildings in Trondheim, Norway. Solar radiation was measured to cause significantly higher temperatures in the air cavity (up to 35–40 degrees) than the outside air temperature. Near the top of the ventilated roof cavity of a sloped roof, temperatures were lower than outside temperatures 42% of the time, compared to 63% at the bottom. For a neighbouring building with a saddle roof, the numbers were 58% and 63% for the south and north side, respectively. Moisture measurements showed that the battens were able to dry quite quickly, going from a moisture level of 37 weight-% to 25% in 38 days.

Tianshu [51] conducted a numerical study of the air flow in air cavities for pitched wooden roofs using computational flow dynamics (CFD). It was found that increasing the width of a cavity from 23 to 48 mm greatly reduced the pressure gradient, friction coefficient, and total loss coefficient. A further increase to 140 mm yielded significantly lower impacts. The shape of the tile battens also greatly influenced the air flow, with a small rounding (r = 4 mm) of the edges reducing the loss coefficient by 20% compared to right-angled battens.

3.4. Building-Integrated Photovoltaics (BIPV)

Sandberg and Mosfegh [55] analysed the air flows and thermal transfer due to thermal buoyancy behind solar panels numerically and experimentally. Laboratory experiments were conducted in Sweden, where thermoelements on the exterior side of the air cavity simulated a solar heating flux of 20–300 W/m². Numerical simulations were conducted with a uniform thermal flux varying between 20 and 500 W/m². The results showed agreement between the laboratory experiments and the simulations and indicated that it is advantageous to use high-emissivity materials for surfaces in air cavities. The analysis revealed the importance of thermal radiation exchange for the heat transfer mechanisms between the air cavity walls. For a thermal flux at or above 200 W/m², almost 30% of the heat was transferred to the otherwise unhated wall through radiation, and from there, it was transferred to the air through convection.

In a further study [56], the same authors investigated governing parameters for air flows generated by the heat transfer between solar panels and air cavities. A laboratory study was conducted, using the variables of inclination angle, the placement of solar panels, and the ratio between air cavity length and width. The results indicate that all the variables greatly influenced the air change ratio and the temperature of the air cavity. The velocity profile was uneven, and the highest airflow velocity did not necessarily occur on the heated side, which indicated a complex airflow pattern. The net effect would be an uneven cooling of the PV module, reducing its electricity generation. The temperature profile showed the highest temperature by the heated wall, the second highest temperature along the opposite wall of the air gap, and the lowest temperature in the middle of the air cavity. The temperature across the entire air gap decreased as the wall/roof angle increased.

In a final study [57], Sandberg and Mosfegh analytically derived expressions for the mass flow rate, air velocity, temperature increase, and location of neutral height (the point where the air pressure in the air cavity is equivalent to the ambient air pressure) in air cavities behind solar panels mounted on vertical façades. The calculations were compared to measured temperature values from previous studies [55,56]. The results indicated a good correlation between analytical expressions and experimental measurements for situations of constrained flow due to the small area of the inlet and/or the outlet compared to the cross-section area of the air cavity. With both ends at the size of the cross-section of the air gap, the agreement was 10–20% but increased at higher wall heights and cavity width ratios. Possible reasons for the discrepancy between the theory and measurements include difficulties with the laboratory measurements and the uncertainty in the many parameters required for a mathematical calculation. Sandberg and Mosfegh's studies comprise the only identified work that concerns the microclimate behind photovoltaic façade elements, unfortunately making comparisons impossible.

3.5. Notable Studies from Outside Nordic Climates

Some studies were identified that directly address the concerns of the present review but that were not included in the main assessment (Table 1) for various reasons. For the sake of reporting the wider international perspective on the microclimate of ventilated air cavities, some notable studies are recounted in this section.

Bouchair [9] presented a theoretical framework for describing cavity airflows in relation to an interior and exterior climate, using the example of a wall built for solar-driven ventilation in an Algerian climate. While "Nordic" parameters such as frost or wind-driven rain were not considered, the equations described may be used to create numerical models of air behaviour in cavities.

In Vancouver, Canada, Tariku and Iffa [58] performed an experimental study assessing the hygrothermal performance of wood frame systems built with different concepts of air cavities (ventilated, vented, no cavity). The findings included that ventilated façades have a vastly higher air change rate; that temperatures behind the façade boards are lower in the wall without an air cavity during winter days; and that the ventilated cavity sees a higher moisture content in the upper section during the wet winter season. This article was not included in the main study as Canada fell outside the scope of Nordic climates; however, the similar temperature conditions of the study site imply that its conclusions can be relevant, nonetheless.

Also in Canada, John Straube and various co-authors [24,25,59,60] examined the role of air cavities in the drying and ventilation of façades. A cavity width as small as 1 mm was found to be enough to ensure drainage. The openings of air cavities were found to play a significant role in the drying potential, they should be as large and unobstructed as possible to ensure drying. Straube also documented that moisture retained in the cavity air and deposited on surfaces was very challenging to simulate in computer models, highlighting the need for field studies and laboratory measurements.

Defo et al. [61] assessed the moisture performance and durability of brick veneer walls in the face of climate change in Canada. The deposition of moisture by wind-driven rain was found to pose a moisture risk on Canada's east coast, and it was found that an air cavity should be accompanied by other measures such as roof overhangs to reduce this risk.

Rahiminejad and Khovalyg [62] reviewed the literature on ventilation rates in ventilated air cavities in façades. A theoretical framework and mathematical models of cavity air flows were presented. The reviewed articles relevant to Nordic climates are all included in the present study.

Van Belleghem et al. [63] modelled the air, heat, and moisture transport in ventilated cavity walls in Belgium, comparing a simplified model in WUFI to a more detailed model. The WUFI model was found to overestimate the ventilation effect on drying and subsequently indicate lower moisture levels than was the case. A lesson learned from this study is that the widely commercially used software may not be sufficient to evaluate the moisture risk in air cavities accurately, presenting a challenge to the building designers who may rely upon it for moisture risk assessment.

Also in Belgium, Langmans and Roels [35] compared four measuring techniques for analysing cavity ventilation behind rainscreen cladding systems. The most suitable method depended on the type of cladding: a vented brick veneer façade displayed air change rates two orders of magnitude lower than ventilated siding, and thus required a pressure gauge to accurately measure the air change rates. The study indicated that measurements of air change rates in air cavities behind brick veneer may not be directly comparable to those behind more open façade claddings.

Marinosci et al. [37] investigated, numerically and experimentally, the thermal behaviour of a rainscreen façade in Italy, featuring an air cavity width of 240 mm. The velocity and temperature distribution of the air in the air cavity was measured and modelled. Correlating values between the model and field measurements suggested that the model is suitable for simulating large air cavities. The study, while thoroughly describing air flows and temperature, did not consider moisture.

4. Discussion

4.1. Air Cavities in Research Literature

An observation recorded early in the literature review was that the literature focusing on brick or rendered façades is much more abundant than the literature on ventilated façades. Even within the latter category, more literature focuses on the cladding itself or the structure behind the wind barrier than on the climate within the air cavity. Many of these articles use assumed climate data for the cavity for calculations or simulations, without performing or citing studies that indicate whether they match the actual conditions. If anything, these articles illustrate the need for the microclimate in air cavities to be investigated further.

The relevant literature recounted in this review originates from a quite small number of researchers at a few research institutions in Norway, Sweden, and Finland. In practice, only six independent field studies and two independent laboratory studies were identified within the relevant scope and climate. Several of the studies lack adequate descriptions of research parameters such as the precise design of the structure, or climate data for the research period, which makes it challenging to replicate or validate the results. The studies are also too different to make it feasible to directly compare their results with each other. For instance, the studies from Finland [44] investigated air cavities with insulation on the exterior side, while the solar panel studies from Sweden [55–57] applied heat sources to the exterior side. Another observed issue is that the studies did not seek to quantify or describe microclimate parameters in the air cavity over time, with a chief focus instead being to compare the performance of different designs.

Wider international literature has been found that addresses these concerns to a greater degree, describing studies that cover hygrothermal conditions in finer detail with several parameters. However, the studies are of limited applicability, as they do not consider temperatures below 0 °C and rarely cover rain penetration. The deposition of liquid water in general remains challenging to simulate in numerical models. It was reported by Van Belleghem et al. [63] that commonly used simulation software tends to underestimate moisture loads, highlighting a challenge in using computer models to investigate high-moisture environments such as air cavities subject to wind-driven rain.

4.2. Climatic Conditions in the Air Cavity

A common observation across many of the field studies was that the temperature of the air cavity will be colder than the ambient air temperature for significant portions of the year, particularly near the inlet opening of the cavity. This creates potential for condensation and moisture damage. There is insufficient data to quantify the moisture risk due to this effect, as the studies are limited to short measurement periods and limited local climates—effectively, relevant measurements have only been conducted in Trondheim in Norway and Lund in Sweden.

Laboratory studies suggest that thermal buoyancy is a quite weak driving force of cavity airflows when compared to wind pressure [49]. The shape of the tile battens has a surprisingly large influence on the airflow resistance of the air cavity of ventilated roofs, with rounded battens reducing the pressure loss coefficient by 20–40% compared to right-angled battens [53]. This effect was observed through both laboratory and CFD studies [51].

4.3. Knowledge Gaps

This study uncovers certain knowledge gaps regarding the microclimate of ventilated cavities in façades and roofs in Nordic climates. There is a general lack of research into the long-term climatic behaviour of air cavities, and the few available studies are not directly comparable to each other. Little systematic research has been conducted that considers all relevant climatic parameters over time, such as the combination of temperature and moisture (in the cavity materials as well as in the air) and how they vary during all seasons. More studies, carried out in a comparable fashion to existing studies, would help address the limited availability of data. Concrete knowledge gaps that have been identified and should be addressed by future studies include:

- To what degree the temperature and relative humidity of the air cavity and correlate with that of the ambient air in long time series, considering the impact of wind, humidity, and rainwater penetration.
- The impact of precipitation and freeze–thaw cycles and how they affect the durability of materials in the building envelope.
- Determining how the air cavity microclimate can be simulated for long-term simulations using exterior climate data in a Nordic climate.
- Determining and describing the climatic loads that materials in the air cavity must withstand to achieve sufficient durability.
- Validating the accuracy of models of humidity conditions in the air cavity, using field or laboratory measurements.

Knowledge about these issues is required to improve the accuracy of simulations and calculations of the durability of materials in ventilated roof and façade systems, increasing the accuracy and reliability of service life predictions.

5. Conclusions

This study showed that the extent of research into the climatic conditions in air cavities in ventilated building roof and façade systems is limited. There is a general lack of comparable studies and quantitative results. The studies are also geographically limited; even though ventilated façade and roof systems are widespread across northern Europe, research into their operating parameters is mainly conducted in Norway and to a limited degree in Sweden. Roofs have generally received more research attention than façades. The studies show how the microclimate of the air cavity may differ from the exterior climate, with the cavity air temperature often being lower than the exterior air temperature. This creates a risk of mould in the air cavity. Thermal buoyancy is also seen to be a relatively weak driving force of air flow in cavities, while the impact of wind is significant.

The implication of the deficiency of studies is that there is currently no good information to use in assessing the durability and service life of materials in air cavities. Industry experience suggests that the principle of two-stage weatherproofing creates durable façade and roof systems, but new types of defects emerge due to a changing climate. It is currently difficult to predict how to improve designs to address these challenges to material durability, as the climatic loads in air cavities are not well studied.

Future work should aim to gather sensor data from buildings spread across different Nordic climate zones over extended durations of time. The relation between the exterior climate—e.g., as described in climate files for simulations—and the climatic conditions of air cavities must also be investigated further. Sensor data may be compared to climate data to evaluate the correlation between the exterior climate and the cavity microclimate, to determine necessary corrections when using climate data to simulate the conditions of ventilated air cavities. Such simulations are essential in evaluating the risk of rot and mould growth in air cavities, and thus the long-term deterioration of components such as the cladding, battens, and wind barrier.

Supplementary Materials: The following supporting information can be downloaded at: https://www.mdpi.com/article/10.3390/buildings12050683/s1, Table S1: Search Strings.

Author Contributions: Conceptualization, S.B.I. and T.K.; methodology, S.B.I., E.A. and T.K.; formal analysis, S.B.I.; investigation, S.B.I.; writing—original draft preparation, S.B.I. and E.A.; writing—review and editing, S.B.I., E.A. and T.K.; supervision, E.A. and T.K.; funding acquisition, T.K. All authors have read and agreed to the published version of the manuscript.

Funding: This research was funded by the Research Council of Norway, grant number 237859.

Institutional Review Board Statement: Not applicable.

Informed Consent Statement: Not applicable.

Acknowledgments: The work is carried out as part of the research project *SFI Klima 2050–Climate adaptation of buildings and infrastructure.* The authors would like to extend a thanks to CAD operator Remy Eik at SINTEF.

Conflicts of Interest: The authors declare no conflict of interest.

References

1. O'Brien, K.; Sygna, L.; Haugen, J.E. Vulnerable or Resilient? A Multi-Scale Assessment of Climate Impacts and Vulnerability in Norway. *Clim. Chang.* **2004**, *64*, 193–225. [CrossRef]
2. Hanssen-Bauer, I.; Drange, H.; Førland, E.J.; Roald, L.A.; Børsheim, K.Y.; Hisdal, H.; Lawrence, D.; Nesje, A.; Sandven, S.; Sorteberg, A. *Klima i Norge 2100-Kunnskapsgrunnlag for Klimatilpasning Oppdatert i 2015*; Norwegian Environmental Agency/Norwegian Climate Service Center: Oslo, Norway, 2015.
3. Lisø, K.R.; Kvande, T. *Klimatilpasning Av Bygninger*, 1st ed.; Sintef Community: Trondheim, Norway, 2007; ISBN 978-82-536-0960-7.
4. Magnussen, K.; Mattsson, J. *Muggsopp i Bygninger. Forekomst og Konsekvenser for Inneklimaet*; Byggforskserien 701.401; SINTEF Community: Trondheim, Norway, 2005.
5. Viitanen, H. Factors Affecting the Development of Biodeterioration in Wooden Constructions. *Mater. Struct.* **1994**, *27*, 483–493. [CrossRef]

6. Thodesen, B.; Kvande, T.; Tajet, H.T.T.; Time, B.; Lohne, J. Adapting Green-Blue Roofs to Nordic Climate. *Nordic Archit. Res.* **2018**, *30*, 99–128.

7. Edvardsen, K.I.; Haug, T.; Ramstad, T. *Trehus. Håndbok 5*; SINTEF Akademisk Forlag: Oslo, Norway, 2018; ISBN 978-825360919.

8. Pourghorban, A.; Asoodeh, H. The Impacts of Advanced Glazing Units on Annual Performance of the Trombe Wall Systems in Cold Climates. *Sustain. Energy Technol. Assess.* **2022**, *51*, 101983. [CrossRef]

9. Bouchair, A. Solar Induced Ventilation in the Algerian and Similar Climates. Ph.D. Thesis, University of Leeds, Leeds, UK, 1989.

10. Kvande, T. *Totrinnstetning Mot Slagregn På Fasader. Luftede Kledninger og Fuger*; Byggforskserien 542.003; SINTEF Community: Trondheim, Norway, 2013.

11. Gaarder, J.E. *Skrå, Luftede Tretak Med Isolerte Takflater-Byggforskserien 525.101*; SINTEF Community: Trondheim, Norway, 2021.

12. Bunkholt, N.S.; Gullbrekken, L.; Time, B.; Kvande, T. Process Induced Building Defects in Norway—Development and Climate Risks. *J. Phys. Conf. Ser.* **2021**, *2069*, 012040. [CrossRef]

13. Bunkholt, N.S.; Time, B.; Kvande, T. *Luftede Kledninger. Anbefalinger for Klimatilpasning*; SINTEF Community: Trondheim, Norway, 2021.

14. Jelle, B.P. Accelerated Climate Ageing of Building Materials, Components and Structures in the Laboratory. *J. Mater. Sci.* **2012**, *47*, 6475–6496. [CrossRef]

15. Hens, H. Modeling the Heat, Air, and Moisture Response of Building Envelopes: What Material Properties Are Needed, How Trustful Are the Predictions? In Proceedings of the 1st Symposium on Heat-Air-Moisture Transport-Measurements on Building Materials, Toronto, ON, Canada, 23 April 2006.

16. Kvande, T.; Bakken, N.; Bergheim, E.; Thue, J.V. Durability of ETICS with Rendering in Norway—Experimental and Field Investigations. *Buildings* **2018**, *8*, 93. [CrossRef]

17. Gullbrekken, L.; Kvande, T.; Jelle, B.P.; Time, B. Norwegian Pitched Roof Defects. *Buildings* **2016**, *6*, 24. [CrossRef]

18. Asphaug, S.K.; Time, B.; Kvande, T. Moisture Accumulation in Building Façades Exposed to Accelerated Artificial Climatic Ageing—A Complementary Analysis to NT Build 495. *Buildings* **2021**, *11*, 568. [CrossRef]

19. Bunkholt, N.S.; Gullbrekken, L.; Kvande, T. Influence of Local Weather Conditions on Ventilation of a Pitched Wooden Roof. *J. Civ. Eng. Archit.* **2020**, *14*, 37–45. [CrossRef]

20. Lexico Dictionaries MICROCLIMATE | Meaning & Definition for UK English | Lexico.Com. In Lexico Dictionaries | English. Available online: https://www.lexico.com/definition/microclimate (accessed on 19 April 2022).

21. Svendsen, S.D. The Principle of One-Stage and Two-Stage Seals. In Proceedings of the International Symposium on Weathertight Joints for Walls, Oslo, Norway, 25 September 1967; pp. 298–301.

22. Birkeland, Ø. The Mechanism of Rain Penetration. In Proceedings of the International Symposium on Weathertight Joints for Walls, Oslo, Norway, 25 September 1967; pp. 33–34.

23. Isaksen, T. Rain Leakage Test on Open Joints in Ventilated Claddings. In Proceedings of the International Symposium on Weathertight Joints for Walls, Oslo, Norway, 25 September 1967; pp. 349–354.

24. Straube, J. The Role of Small Gaps Behind Wall Claddings on Drainage and Drying. In Proceedings of the 11th Canadian Conference on Building Science & Technology, Banff, AB, Canada, 22 March 2007.

25. Straube, J.; Smegal, J. *Modeled and Measured Drainage, Storage and Drying Behind Cladding Systems*; Building Science Corporation: Somerville, MA, USA, 2009.

26. Arksey, H.; O'Malley, L. Scoping Studies: Towards a Methodological Framework. *Int. J. Soc. Res. Methodol.* **2005**, *8*, 19–32. [CrossRef]

27. O'Brien, A.M.; Mc Guckin, C. The Systematic Literature Review Method: Trials and Tribulations of Electronic Database Searching at Doctoral Level. In *SAGE Research Methods Cases*; SAGE Publishing Ltd.: New York, NY, USA, 2016. [CrossRef]

28. Johansen, K.S. Internal Rain Gutter for BIPV Roof. Dimensioning the Internal Rain Gutter for ZEB Laboratory's BIPV Roof. Master's Thesis, Norwegian University of Science and Technology, Trondheim, Norway, 2019.

29. Wohlin, C. Guidelines for Snowballing in Systematic Literature Studies and a Replication in Software Engineering. In *Proceedings of the 18th International Conference on Evaluation and Assessment in Software Engineering—EASE '14*; ACM Press: London, UK, 2014; pp. 1–10.

30. Gobakken, L.R.; Høibø, O.A.; Solheim, H. Factors Influencing Surface Mould Growth on Wooden Claddings Exposed Outdoors. *Wood Mater. Sci. Eng.* **2010**, *5*, 1–12. [CrossRef]

31. Janssen, H.; Blocken, B.; Carmeliet, J. Conservative Modelling of the Moisture and Heat Transfer in Building Components under Atmospheric Excitation. *Int. J. Heat Mass Transf.* **2007**, *50*, 1128–1140. [CrossRef]

32. Gupta, B.S.; Jelle, B.P.; Hovde, P.J.; Gao, T. Wood Coating Failures against Natural and Accelerated Climates. *Proc. Inst. Civ. Eng.-Constr. Mater.* **2015**, *168*, 3–15. [CrossRef]

33. Geving, S.; Holme, J. Vapour Retarders in Wood Frame Walls and Their Effect on the Drying Capability. *Front. Archit. Res.* **2013**, *2*, 42–49. [CrossRef]

34. Lie, S.K.; Thiis, T.K.; Vestøl, G.I.; Høibø, O.; Gobakken, L.R. Can Existing Mould Growth Models Be Used to Predict Mould Growth on Wooden Claddings Exposed to Transient Wetting? *Build. Environ.* **2019**, *152*, 192–203. [CrossRef]

35. Langmans, J.; Roels, S. Experimental Analysis of Cavity Ventilation behind Rainscreen Cladding Systems: A Comparison of Four Measuring Techniques. *Build. Environ.* **2015**, *87*, 177–192. [CrossRef]

36. Brischke, C.; Thelandersson, S. Modelling the Outdoor Performance of Wood Products—A Review on Existing Approaches. *Constr. Build. Mater.* **2014**, *66*, 384–397. [CrossRef]
37. Marinosci, C.; Strachan, P.A.; Semprini, G.; Morini, G.L. Empirical Validation and Modelling of a Naturally Ventilated Rainscreen Facade Building. *Energy Build.* **2011**, *43*, 853–863. [CrossRef]
38. Falk, J.; Sandin, K. Ventilated Rainscreen Cladding: Measurements of Cavity Air Velocities, Estimation of Air Change Rates and Evaluation of Driving Forces. *Build. Environ.* **2013**, *59*, 164–176. [CrossRef]
39. Falk, J.; Sandin, K. Ventilated Rainscreen Cladding: A Study of the Ventilation Drying Process. *Build. Environ.* **2013**, *60*, 173–184. [CrossRef]
40. Falk, J.; Molnár, M.; Larsson, O. Investigation of a Simple Approach to Predict Rainscreen Wall Ventilation Rates for Hygrothermal Simulation Purposes. *Build. Environ.* **2014**, *73*, 88–96. [CrossRef]
41. Uvsløkk, S. The Importance of Wind Barriers for Insulated Timber Frame Constructions. *J. Therm. Insul. Build. Envel.* **1996**, *20*, 40–62. [CrossRef]
42. Nore, K. Hygrothermal Performance of Ventilated Wooden Cladding. Ph.D. Thesis, Norwegian University of Science and Technology, Trondheim, Norway, 2009.
43. Nore, K.; Blocken, B.; Thue, J.V. On CFD Simulation of Wind-Induced Airflow in Narrow Ventilated Facade Cavities: Coupled and Decoupled Simulations and Modelling Limitations. *Build. Environ.* **2010**, *45*, 1834–1846. [CrossRef]
44. Viljanen, K.; Lü, X.; Puttonen, J. Factors Affecting the Performance of Ventilation Cavities in Highly Insulated Assemblies. *J. Build. Phys.* **2021**, *45*, 67–110. [CrossRef]
45. Gullbrekken, L.; Uvsløkk, S.; Geving, S.; Kvande, T. Local Loss Coefficients inside Air Cavity of Ventilated Pitched Roofs. *J. Build. Phys.* **2018**, *42*, 197–219. [CrossRef]
46. Hansen, E. Luftstrømning i skrå tretak-Eksperimentelle undersøkelser og numeriske beregninger. Master's Thesis, Norwegian University of Science and Technology, Trondheim, Norway, 2016.
47. Eggen, M.G.; Røer, O.V. Lufting av Skrå Tretak-Trykktap ved Ulike Luftespalteutforminger. Master's Thesis, Norwegian University of Science and Technology, Trondheim, Norway, 2018.
48. Bunkholt, N.S.; Säwén, T.; Stockhaus, M.; Kvande, T.; Gullbrekken, L.; Wahlgren, P.; Lohne, J. Experimental Study of Thermal Buoyancy in the Cavity of Ventilated Roofs. *Buildings* **2020**, *10*, 8. [CrossRef]
49. Bunkholt, N.S. Eksperimentell studie av termisk oppdrift i tak med luftet tekning. Master's Thesis, Norwegian University of Science and Technology, Trondheim, Norway, 2019.
50. Säwén, T.; Stockhaus, M.; Hagentoft, C.E.; Bunkholt, N.S.; Wahlgren, P. Model of Thermal Buoyancy in Cavity-Ventilated Roof Constructions. *J. Build. Phys.* **2021**, *45*, 413–431. [CrossRef]
51. Tianshu, L. Numerical Study of Air Flow in Air Cavities for Pitched Wooden Roofs. Master's Thesis, Norwegian University of Science and Technology, Trondheim, Norway, 2020.
52. Mundt-Petersen, S.O.; Harderup, L.E. Predicting Hygrothermal Performance in Cold Roofs Using a 1D Transient Heat and Moisture Calculation Tool. *Build. Environ.* **2015**, *90*, 215–231. [CrossRef]
53. Gullbrekken, L.; Kvande, T.; Time, B. Ventilated Wooden Roofs: Influence of Local Weather Conditions—Measurements. *Energy Procedia* **2017**, *132*, 777–782. [CrossRef]
54. Rønningen, E.S. Feltstudie Av Klimatiske Forhold i Luftespalter Bak Kledning Og Taktekning. Master's Thesis, Norwegian University of Science and Technology, Trondheim, Norway, 2020.
55. Sandberg, M.; Moshfegh, B. Flow and Heat Transfer in the Air Gap behind Photovoltaic Panels. *Renew. Sustain. Energy Rev.* **1998**, *2*, 287–301. [CrossRef]
56. Sandberg, M.; Moshfegh, B. Ventilated Solar Roof Air Flow and Heat Tranfer Investigation. *Renew. Energy* **1998**, *15*, 287–292. [CrossRef]
57. Sandberg, M.; Moshfegh, B. Buoyancy-Induced Air Flow in Photovoltaic Facades—Effect of Geometry of the Air Gap and Location of Solar Cell Modules. *Build. Environ.* **2002**, *37*, 211–218. [CrossRef]
58. Tariku, F.; Iffa, E. Empirical Model for Cavity Ventilation and Hygrothermal Performance Assessment of Wood Frame Wall Systems: Experimental Study. *Build. Environ.* **2019**, *157*, 112–126. [CrossRef]
59. Straube, J.; van Straaten, R.; Burnett, E. Field Studies of Ventilation Drying. In Proceedings of the 9th International Conference on Performance of Exterior Envelopes of Whole Buildings (Buildings IX), Clearwater Beach, FL, USA, 5–10 December 2004.
60. Straube, J.F.; Finch, G. *Ventilated Wall Claddings: Review, Field Performance, and Hygrothermal Modelling*; Research Report-0906; Building Science Corporation: Westford, MA, USA, 2009.
61. Defo, M.; Lacasse, M.A.; Wang, L. Effects of Climate Change on the Moisture Performance and Durability of Brick Veneer Walls of Wood Frame Construction in Canada. *J. Phys. Conf. Ser.* **2021**, *2069*, 012063. [CrossRef]
62. Rahiminejad, M.; Khovalyg, D. Review on Ventilation Rates in the Ventilated Air-Spaces behind Common Wall Assemblies with External Cladding. *Build. Environ.* **2021**, *190*, 107538. [CrossRef]
63. Van Belleghem, M.; Steeman, M.; Janssens, A.; De Paepe, M. Heat, Air and Moisture Transport Modelling in Ventilated Cavity Walls. *J. Build. Phys.* **2015**, *38*, 317–349. [CrossRef]

 *buildings*

Article

Structural Performance Assessment of Innovative Hollow Cellular Panels for Modular Flooring System

Keerthana John [1,*], Sherin Rahman [1], Bidur Kafle [1], Matthias Weiss [2], Klaus Hansen [3], Mohamed Elchalakani [4], Nilupa Udawatta [5], M. Reza Hosseini [5] and Riyadh Al-Ameri [1]

1 School of Engineering, Deakin University, Geelong, VIC 3217, Australia; skrahman@deakin.edu.au (S.R.); bidur.kafle@deakin.edu.au (B.K.); r.alameri@deakin.edu.au (R.A.-A.)
2 Institute for Frontier Materials, Deakin University, Geelong, VIC 3217, Australia; matthias.weiss@deakin.edu.au
3 L2U Pty. Ltd., Perth, WA 6017, Australia; klaus@nksom.com
4 Civil, Environmental and Mining Engineering, The University of Western Australia, Perth, WA 6009, Australia; mohamed.elchalakani@uwa.edu.au
5 School of Architecture & Built Environment, Deakin University, Geelong, VIC 3217, Australia; nilupa.udawatta@deakin.edu.au (N.U.); reza.hosseini@deakin.edu.au (M.R.H.)
* Correspondence: johnke@deakin.edu.au

Abstract: Lightweight modular construction has become an increasing need to meet the housing requirements around the world today. The benefits of modular construction ranging from rapid production, consistency in quality, sustainability, and ease of use have widened the scope for the construction of residential, commercial, and even emergency preparedness facilities. This study introduces novel floor panels that can be flat-packed and built into modular housing components on-site with minimal labour and assistance. The flooring system uses hollow cellular panels made of various configurations of trapezoidal steel sheets. The structural performance of three different configurations of these hollow flooring systems as a modular component is presented in this study by analysing the failure modes, load-displacement parameters, and strain behaviour. The study confirms significant advantages of the proposed hollow floor systems, with multi-cells reporting higher load-carrying capacity. The hollow flooring system performed well in terms of structural performance and ease in fabrication as opposed to the conventional formworks and commercial temporary flooring systems. The proposed flooring system promises efficient application as working platforms or formworks in temporary infrastructural facilities and emergency construction activities.

Keywords: cellular flooring; modular construction; profiled sheets; structural performance

Citation: John, K.; Rahman, S.; Kafle, B.; Weiss, M.; Hansen, K.; Elchalakani, M.; Udawatta, N.; Hosseini, M.R.; Al-Ameri, R. Structural Performance Assessment of Innovative Hollow Cellular Panels for Modular Flooring System. *Buildings* **2022**, *12*, 57. https://doi.org/10.3390/buildings12010057

Academic Editor: Jian-Guo Dai

Received: 14 December 2021
Accepted: 5 January 2022
Published: 6 January 2022

Publisher's Note: MDPI stays neutral with regard to jurisdictional claims in published maps and institutional affiliations.

1. Introduction

The construction of an emergency housing system is still one of the priorities in the 21st century for many humanitarian and government agencies due to various natural disasters and civilian crises. It is estimated that approximately 260 million people migrate to different parts of the globe, creating increased demand in the housing sector [1]. With an anticipated increase in these numbers, it is high time for the construction sector to fast track research on affordable, fast-paced, modular and adaptable housing systems to meet the needs of the future generation [1,2]. The construction sector has strived to address such increased demand, and hence the focus is to develop thoughtful innovations in construction, method, technology and equipment [2,3]. The adoption of offsite construction is one such innovation in the construction industry that enables the workers and machinery to work in a controlled environment. Offsite construction has also increased productivity, quality and safety due to reduced external influences on the construction stage [4].

Offsite construction is the future of the building industry, where construction processes resemble manufacturing procedures and construction sites are used only for assembling modular building components [5]. In this construction method, the whole structure can be assembled with prefabricated components that can be transported and installed on-site [6,7]. The prefabricated modules can include beams, columns, slabs, stairs, and panel elements of various construction materials, including concrete, steel, timber, or composite materials [6,8]. This method has several benefits such as improved quality, enhanced structural reliability and productivity. It also reduces the construction time, labour and wastage because of the controlled environment under which they are manufactured [9,10]. The modular construction process incorporates design optimisation, and there is an opportunity to reuse the building components effectively [11]. Thus, the modular construction method aligns with circular economy principles and ensures that all building components fulfil structural performance requirements, building codes, and standards [12,13].

Modular construction is also highly conducive to supporting sustainability initiatives. Considering these aspects, lightweight steel modular units are becoming popular for constructing emergency facilities due to their acceptable structural performance, fire resistance and lightweight characteristics [14]. These modular steel components have shown enhanced strength and rigidity, as evidenced by various studies [9,15]. However, there is a need of designing effective connections for assembling the modular components. Steel modular structures with easy connectors are a good alternative for immediate response for the industry [2,8]. Despite several advantages of modular construction over the conventional construction process, the structures industry still rely heavily on the traditional on-site construction methods [4,9]. There is a lack of lightweight, integrated, and reconfigurable building components, especially floor systems [16]. A review of the existing literature suggests enormous potential for investigating the structural efficacy of hollow flooring systems made of lightweight corrugated steel sheets [17]. Some studies indicate that the use of profiled steel sheets is increasing to provide quality and affordable housing in the form of self-supporting roofs and other applications owing to their structural capabilities without any support [11,18]. Experimental investigation of profiled steel sheets for their constructability in various applications such as composite flooring systems have shown encouraging results [19,20].

This experimental research develops a novel modular flooring system with L2U Group Pty Ltd. (L2U) from Perth, Australia, FormFlow [21], Geelong, Australia and Deakin University, from Geelong, Australia. This study specifically focused on developing floor hollow floor decks from simple trapezoidal corrugated sheets for temporary structural applications in the emergency housing sector. The prototypes of the flooring panel were made using flat-packed corrugated sheets supplied by L2U Group Pvt. Ltd., Perth, Australia which were then assembled in the Deakin University laboratory facility in Geelong. This paper presents the test results of the structural performance of three different cell configurations of the novel floor panels. Detailed analysis of the load-displacement behaviour, failure modes and strain data of various panel configurations and their assessment for suitability in temporary floor construction is presented.

2. Materials and Methods

The test program fabricates the innovative cellular panels from corrugated profile sheets. It tests their performance as a temporary floor panel for emergency housing and as a working platform for other construction applications. The fabrication of the cellular panel from the flat-packed profile sheets was conducted following the U.S. Patent 8,539,730 B2 [22]. In total, four different hollow cell configurations were identified to investigate the performance in various arrangements, two of them are single cell, and two were multi-cell arrangements.

Hence, the current proposal introduces and tests a method for simple flooring made using an overlapping pattern of locally available corrugated sheets [21]. It enables the construction of a flooring system even in an irregular terrain ruling out the need for initial site preparation [21]. Moreover, this type of flooring does not require skilled labour and can be used with ease, especially for emergency structures such as in refugee shelters and temporary housing platforms.

2.1. Fabrication of the Innovative Hollow Cellular Panels

The hollow panels made of simple trapezoidal profiled corrugated sheets (Figure 1a) were fabricated by L2U Pty. Ltd., a start-up company in Western Australia with the assistance of FormFlow for prototyping, design, and logistics. FormFlow is a leading Australian developer of advanced steel building solutions and recently patented a world-first technology to produce sharp 90 degree bends in corrugated sheets; the product has been commercialised in Australia as Lysaght CUSTOMFLOW [23]. A simple trapezoidal-corrugated steel strip commonly employed in roof or wall cladding throughout Australia has been adopted by L2U for the new system. The flat-packed sheets provided for this study were 1200 mm (L) × 600 mm (W) sizes with 0.42 mm sheet thickness. All the materials used in the study were provided by our industry partner—L2U group Pvt. Ltd., Perth, Australia. These are commercially available as Lysaght Trimdek, which is a commonly used trapezoidal corrugated steel sheet. The corrugated steel used is coated with ZINCALUME (aluminium/zinc/magnesium alloy) which complies with AS1397:2011. The sheet offers yield strength of 550 MPa with a coating mass of 125 g/m^2. These details are provided by the manufacturer for the purpose of our study [24].

The sheets are then bent along the axis of the slots at 90°, as shown in Figure 1a. Custom slots are provided along the longitudinal direction of the sheet (Figure 1b,c) at the required distance using the Computerised Numerical Control or CNC cutting technique for ease of bending at ends. The flat-packed corrugated sheets are bent halfway into a 'C' shaped panel section, which can be laid over the adjacent panels until the desired span is achieved and closed at the end section by fasteners or bolts. Plates can also be included at the end of the middle section of the panels to provide improved structural integrity (Figure 1d). Multiple sheets for longer spans can also be fabricated and connected this way. When folded over the adjacent cell, an overlap is formed at the folds, and this configuration results in the cells being significantly robust, as shown in Figure 1e. The end plates connect the top and bottom skins of the steel sheets to form a complete cell.

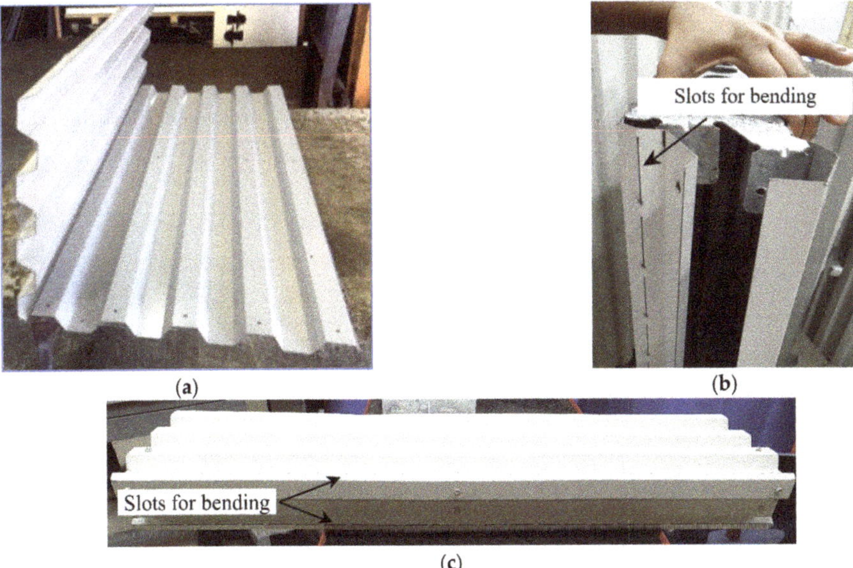

(a)

(b)

(c)

Figure 1. *Cont.*

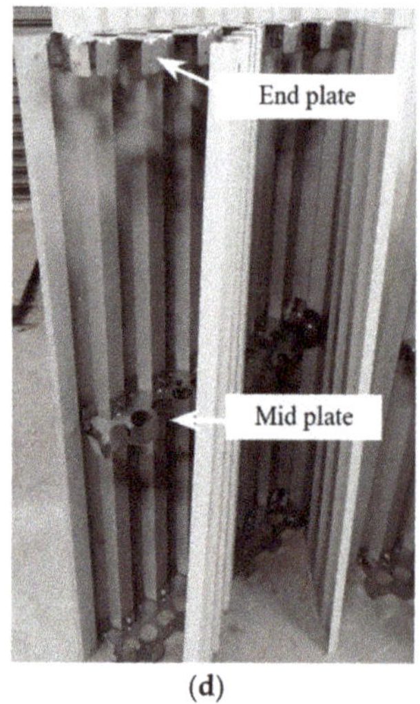

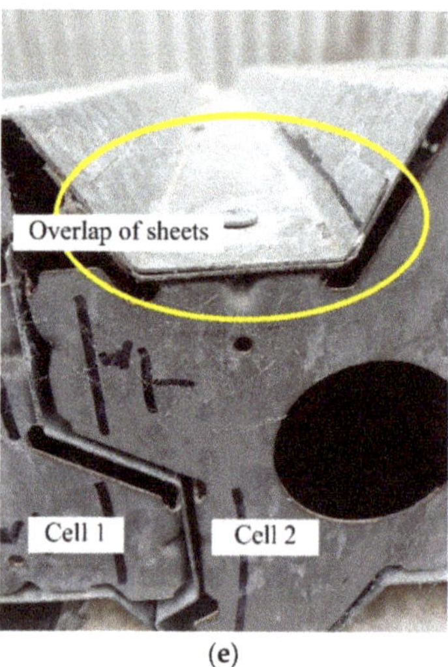

(**d**)

(**e**)

Figure 1. Prototype of hollow cellular panel. (**a**) Bent 'L' shaped corrugated sheet. (**b**) Slots for bending at open end. (**c**) Slots for bending at corrugation edges. (**d**) Open panel with mid plates (**e**) Overlap of sheets of a multi-cell panel.

2.2. Panel Details

This study investigates three different panel configurations: a single cell, a double cell, and a triple cell (shown in Figure 2). The details of the test specimens' dimensions, panel types, and configurations are provided in Table 1.

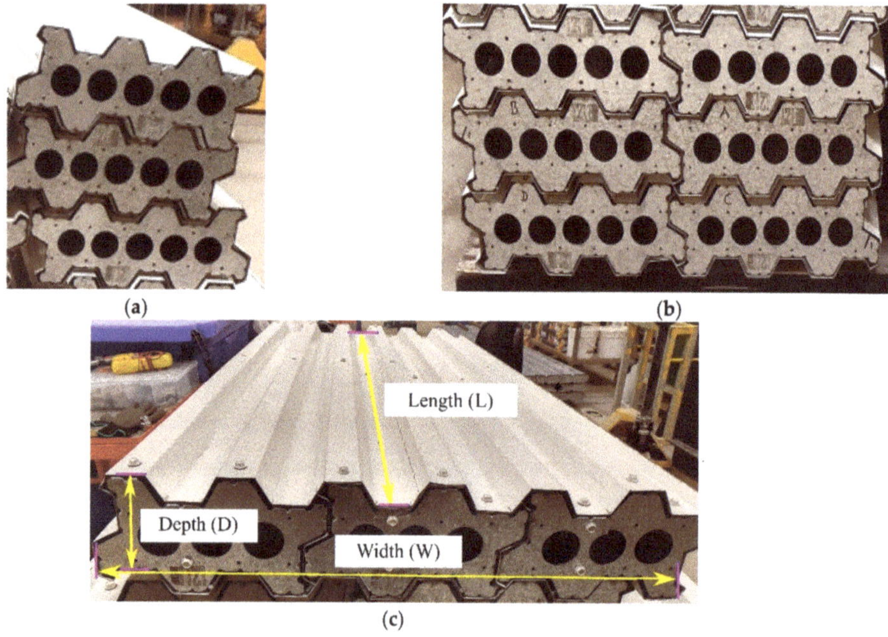

(**a**)

(**b**)

(**c**)

Figure 2. (**a**) Single-cell panels. (**b**) Double cell panels. (**c**) Triple Cell Panels with annotations of dimensions used.

Table 1. Details of Test Specimens.

Panel Type	Panel ID	Panel Dimensions (L × W × D) mm	Width of Individual Cell (W) mm	Number of Panels Tested
Single cell (with mid-plate)	SC-P	1200 × 300 × 100	300	2
Single cell (without mid-plate)	SC	1200 × 300 × 100	300	1
Double cells (without mid-plate)	DC	1200 × 600 × 100	300	2
Triple cells (without mid-plate)	TC	1200 × 600 × 100	200	2
Total number of panels				7

Out of the above test specimens, a couple of single cells were provided with a mid-plate to assess its role in the structural behaviour of the panel for initial investigation. The mid-plate position is shown in Figure 1d. This plate is made of steel, cut to the panel's shape, and is provided at the centre of the panels at 600 mm as a support element. Multi-cell panels were tested without mid-plates to achieve a complete hollow cellular system. It was assumed that the overlap of sheets in multi-cell panels would enhance the structural efficiency eliminating the need for mid-plates. Thus, the multi-cell panels were kept at a constant total width of 600 mm with an individual cell width of 300 mm for double cell panels and 200 mm for the triple cell panels to assess the structural performance of the overlapping sheet system. Another aim of testing multi-cell panels with different cell widths was to observe the role of various cell configurations in their overall structural behaviour as a flooring system.

2.3. Test Set-Up

A four-point bending arrangement was followed to test the hollow cellular panels as depicted in Figure 3a,b. The panels were simply supported at an effective span of 1000 mm. The load was applied at the mid-span through a hydraulic piston onto roller supports at 334 mm spacing from each other. The test was displacement controlled at the rate of 2.5 mm/min. A laser transducer was placed underneath the panel at the centre of the panel to measure the mid-point displacement. Electrical resistance strain gauges were positioned at the top and bottom of the panel. Three strain gauges were positioned at the bottom of the panel: SG1 and SG3 below the load points and SG2 at the centre of the panel. At the same time, SG4 was placed at the centre of the top skin of the panel (Figure 3d). A data acquisition system connected to the load frame was programmed to capture the applied load, displacement, and strain data. The panels were supported by timber blocks positioned in the panels and above the supports, as shown in Figure 3c. It was done to prevent the primary crushing of corrugations when the load is applied and to ensure an appropriate failure mode.

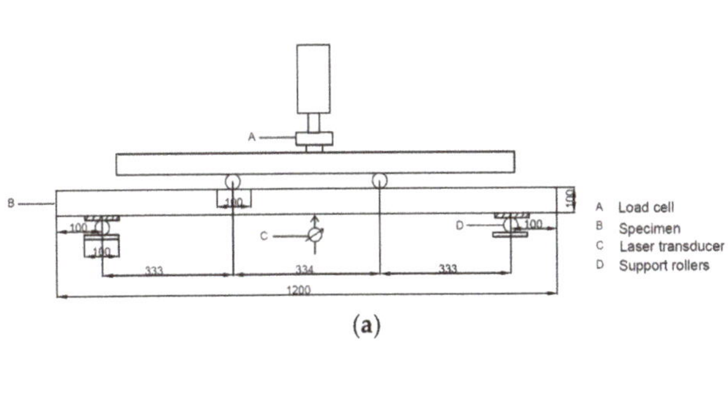

(a)

(b)

Figure 3. *Cont.*

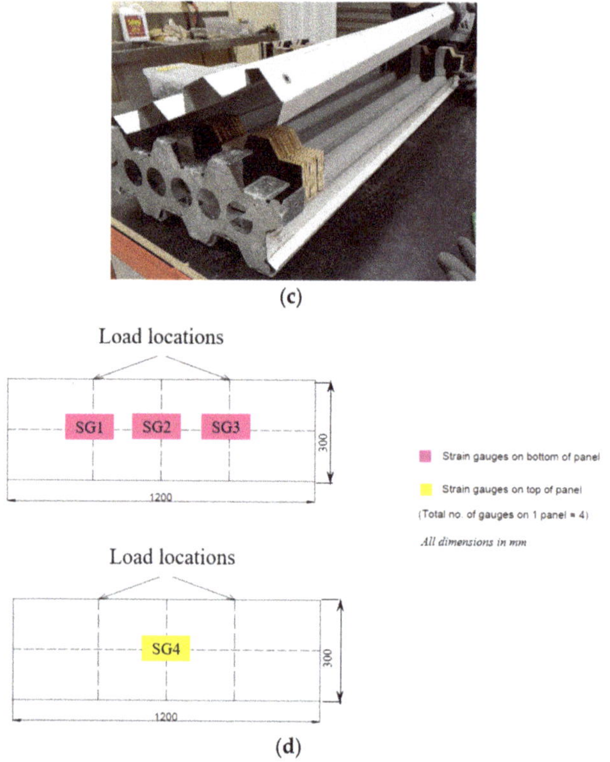

Figure 3. (**a**) Test set-up for panels. (**b**) Test arrangement of the single-cell panel. (**c**) Open panel with timber blocks inserted over end support regions. (**d**) Position of strain gauges.

3. Results and Discussion

3.1. Load-Displacement Behaviour

3.1.1. Single-Cell Panels

Both single-cell panels with mid-plate (SC-P) attained an average ultimate load of 6.34 kN, which is 76% higher than the panel without mid-plate (SC) which had an ultimate load of 3.60 kN. It suggests that the mid-plate contributes to a significant load-carrying capacity of the panels. The mid-plate also mobilises the contribution of the lower sheet in load-carrying capacity at the early stages of loading. In the panels without a mid-plate, the lower sheet will only take part in load resistance at the later stages of loading due to the lack of connectivity with the top plate. It is seen from Figure 4 that all single-cell panels exhibit excellent ductile behaviour.

In SC-P panels, the load is maintained due to mid-plate and gradually decreases when the skins buckled over the mid-plate. However, the load does not undergo any significant reduction in the SC panel and continues steadily to carry the load until the test is stopped at 90 mm displacement. The panel has already suffered local buckling at this stage, and high displacements were attained. Hence irrespective of the load-carrying capacity, the test was stopped at this stage. The top skin of the panel acted independently due to the absence of a mid-plate to transfer the load to the bottom skin. Once it touched the bottom skin, the ultimate load-carrying capacity was maintained due to double skin action, seen from the load-displacement graph in Figure 4 and Table 2.

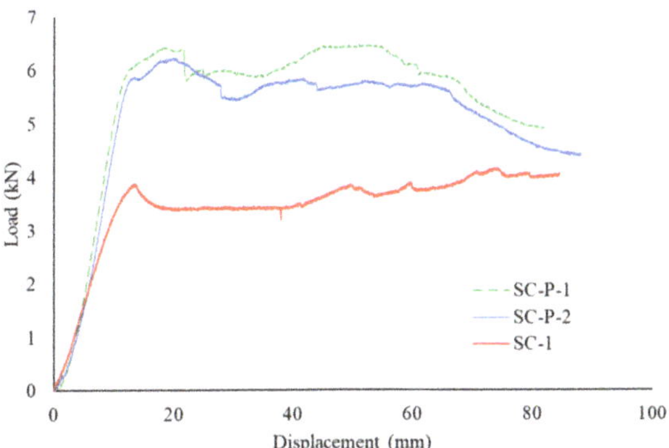

Figure 4. Load vs. displacement of single-cell panels.

Table 2. Results of single cell panel tests.

Panel	Ultimate Load (kN)	Displacement—Top Skin (mm)	Displacement—Bottom Skin (mm)
SC-P-1	6.23	20.10	19.98
SC-P-2	6.46	21.13	21.15
Average	6.34	20.61	20.56
SC	3.60	15.30	0.91

To further investigate the role of the double skin action in the SC panel, the load-displacement relationship of the top and bottom skins is plotted as shown in Figure 5. Double skin action of the panels here is defined as the load carrying capacities created by the top and bottom skins when acting together.

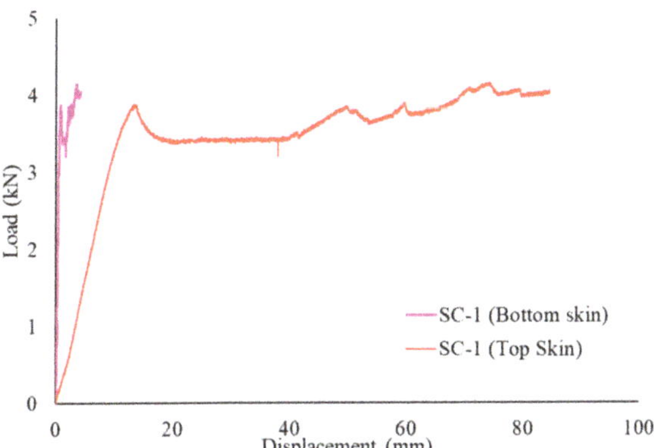

Figure 5. Load vs. displacement of top and bottom skins in SC panel.

Displacement on the top skin is significantly high, whereas the bottom skin started to deflect later without any noticeable amount compared with the top skin. During the test, the displacement in the bottom skin was initiated only after the top skin touched the bottom skin. This observation is evidenced by the bottom and top skin displacement at ultimate loads as 0.91 mm and 15.30 mm, respectively (Table 2). It implies that the mid-plate contributed to distributing loads and displacements between the top and bottom skins.

3.1.2. Multi-Cell Panels

The load-displacement behaviours of both double and triple cells are presented in Figure 6.

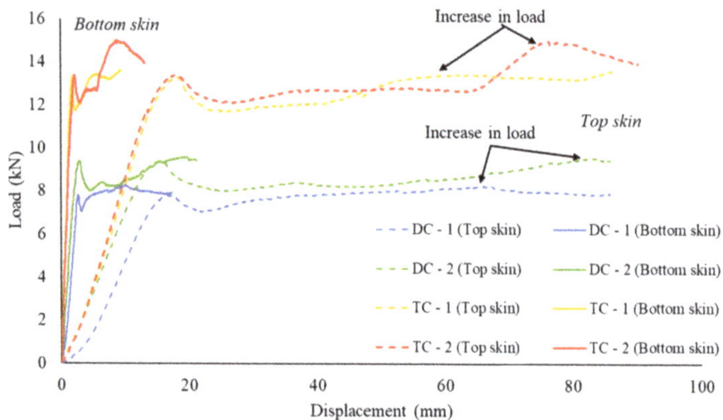

Figure 6. Load vs. displacement of multi-cell panels.

It is evident from Figure 6 that the triple cell panels attained a significantly higher average ultimate load of 13.34 kN compared to double cell panels with an average ultimate load of 8.6 kN. There is around a 55% increase in the ultimate load-carrying capacity from the double to triple cells owing to the additional partitioning steel walls in the 600 mm cross-sections of the triple cell panels. Hence it is evident that any addition of cells multiplies the load capacity due to additional overlap of sheets. Like single-cell panels, excellent ductile behaviour was seen for all panels, as shown in Figure 6. Ultimate load is achieved at a lower displacement and a steady load without considerable load reduction. When the top skin touches the bottom skin, the maximum load is again achieved as indicated in the graphs as increased load, and the test is terminated at 90 mm displacement. The displacement measured on the bottom skin indicates that this skin is only activated at the later stages and undergoes minor displacement, similar to the single-cell panel (SC). The test results are compiled in Table 3 below.

Table 3. Results of multi-cell panel tests.

Panel	Panel Dimensions (mm)	Width of Individual Cell (mm)	Number of Cells	Ultimate Load (kN)	Displacement—Top Skin (mm)	Displacement—Bottom Skin (mm)
DC—1	1200 × 600 × 100	300	2	7.81	16.46	2.61
DC—2	1200 × 600 × 100	300	2	9.38	15.47	2.86
	Average			8.59	15.96	2.73
TC—1	1200 × 600 × 100	200	3	13.30	18.03	1.78
TC—2	1200 × 600 × 100	200	3	13.38	17.77	2.12
	Average			13.34	17.90	1.95

3.1.3. Summary of Load-Displacement Behaviour

Adopting single, double, and triple cell configurations helped in understanding the load distribution in the different panels. The strength multiplication in the double cells of 600 mm width from the single cells of 300 mm width confirm that the one-way slab action is maintained for all the slab configurations used in this study. Furthermore, when the mid-plates are provided for the single-celled panel configuration (SC-P), the load-carrying capacity increases by 76% compared with the single-celled panels without mid-plate (SC).

From the load-displacement curves, it is evident that any addition of cell numbers creates a more stiffened panels, thereby multiplying the load-carrying capacity of the proposed innovative flooring system. The strength improvement is therefore due to the overlapping action in the multi-celled panels. In the current test program, the triple celled panels achieved 55% higher ultimate loads than the double celled panels even though both had the same overall dimensions of 1200 mm (L) × 600 mm (W) and 100 mm depth (D).

Table 4 is shown for further comparison of the performance of the 300 mm wide single cells and 600 mm wide double cells.

Table 4. Comparison of average load vs displacement values for single cell and double cell panels.

Panel	Panel Dimensions (mm)	Width of Individual Cell (mm)	No. of Cells	Ultimate Load (kN)	Displacement—Top Skin (mm)	Displacement—Bottom Skin (mm)
Single Cell (Without mid-plate)	1200 × 300 × 100	300	1	3.60	15.30	0.91
Single cell (With mid-plate)	1200 × 300 × 100	300	1	6.34	20.61	20.56
Double Cell	1200 × 600 × 100	300	2	8.59	15.96	2.73

From Table 4 it is clear that on multiplying the cell numbers, there is significant increase in the load capacity, however maintaining the one-way load distribution irrespective of the doubling of cells. A 35.50% increase is seen between single cells with mid-plates and double cells without mid-plate, however, the panel with mid-plates reported higher displacements. The mid-plates are therefore beneficial for higher load carrying applications and ensures complete utilization of the top and bottom skins of the panels for load transfer.

To further investigate the effect of overlapping on multi cell panels with same overall dimensions, ultimate load, and displacement on two types of multi cells (double and triple) panels have been compared and presented in Table 5. The triple cell panel achieved 55% more ultimate load compared with double cell panel with similar overall dimensions, confirming the benefits of additional partitioning walls. On the other hand, the displacements at top and bottom skins are comparable on both double and triple cells, indicating that triple cell can resist more loads without significant increase in displacement.

Table 5. Comparison of average load vs. displacement values for double cell and triple cell panels.

Panel	Panel Dimensions (mm)	Width of Individual Cell (mm)	No. of Cells	Ultimate Load (kN)	Displacement—Top Skin (mm)	Displacement—Bottom Skin (mm)
Double Cell	1200 × 600 × 100	300	2	8.59	15.96	2.73
Triple cell	1200 × 600 × 100	300	3	13.34	17.90	1.95

The results from these tests confirm that the current design of panels ensures structural adequacy in carrying construction loads. Thus, it implies that the proposed panel's configurations are adequate for the application in temporary structures for formwork, emergency shelters, etc.

3.2. Failure Mode

3.2.1. Single-Cell Panels

Panels with Mid-Plate

The mid-plate in single-cell panels, SC-P, provided an excellent resistance to load. Flexural failure of the panel (Figure 7a), local buckling around the mid-plate (Figure 7b) and below the loading point (Figure 7c) were observed. The slots provided for bending of the sheets also suffered tearing at these critical points. The failure of slots can be attributed to compression of the top skin of the panel over the mid-plate. This behaviour was noticed towards the end of the test when large displacement values beyond 50 mm were already attained. The top skin of the panels also began to separate from the endplates due to these high displacements, as shown in Figure 7d. However, it was observed that the mid-plate was acting as a medium to transfer the load from top skin to bottom skin of the panel and hence the full capacity of the panel was exploited.

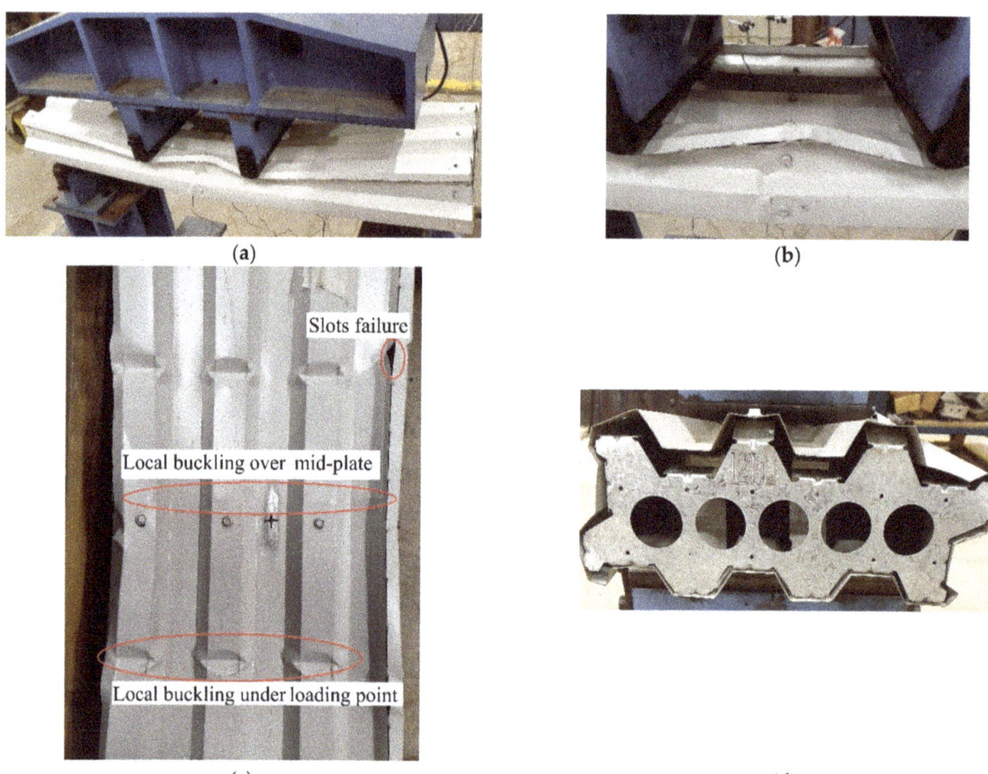

Figure 7. (**a**) Failed SC-P panel. (**b**) Buckling over mid-plate in SC-P. (**c**) Failure points on SC-P panel (**d**) Separation of panel skin from endplates.

Panels without Mid-Plate

The panel without the mid-plate also failed by flexure (Figure 8a). However, local buckling was noticed only below the load points. It is seen from Figure 8b that only the top skin of the panel suffered failure while the bottom skin remained significantly undamaged. Similar to panels with mid-plates, separation of the skins from the endplate was noticed at higher displacement beyond 50 mm.

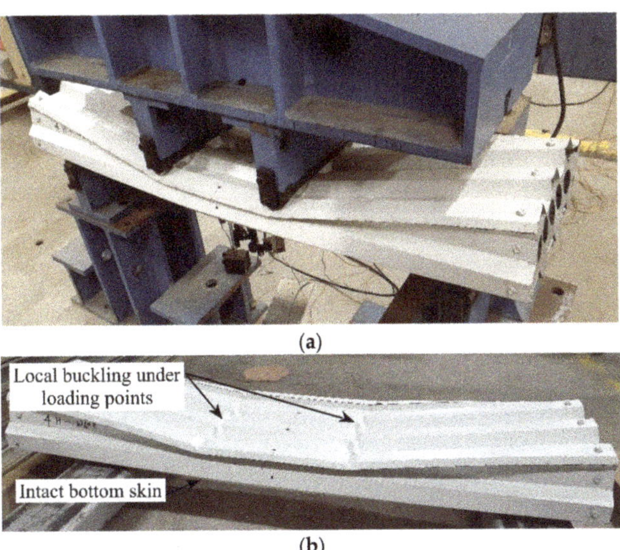

(a)

(b)

Figure 8. (**a**) Failure of SC panel. (**b**) Failed SC panel with the intact bottom skin.

3.2.2. Multi-Cell Panels

The failure modes of both double and triple cell panels were very similar, and both underwent flexural failure (Figure 9a). All multi-cell panels experienced the same failure behaviour. Local buckling was noticed only under the load points, as shown in Figure 9b, without any failure of the slots. The top and bottom skin acted independently, and hence the bottom skin was not affected by any local failure and remained undamaged (Figure 9c). However, the bottom skins flattened themselves (Figure 9d) at the slots, which deformed the panel along the centre. It commenced when the top skin began touching the bottom skin due to displacement. The overlap in one of the double cell panels suffered separation at the centre of the panel due to high displacement (Figure 9e). However, this was not observed in the case of the triple cells. Consistent with single-cell panels, the top skin began detaching from the endplate when higher displacements of over 50 mm were initiated.

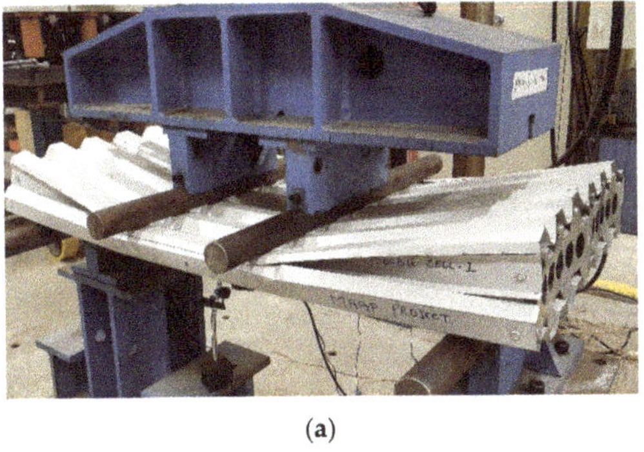

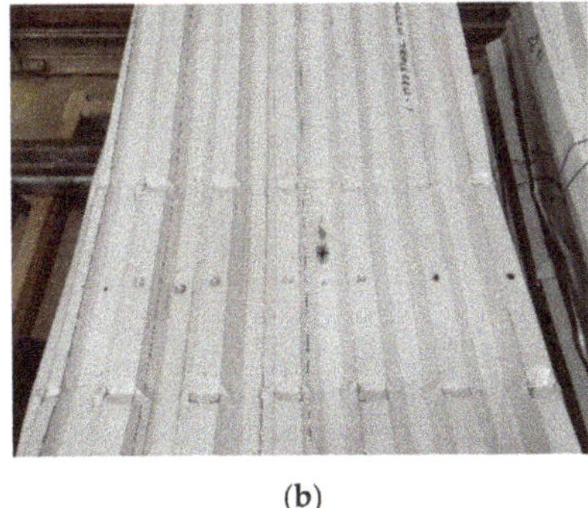

(a)

(b)

Figure 9. *Cont.*

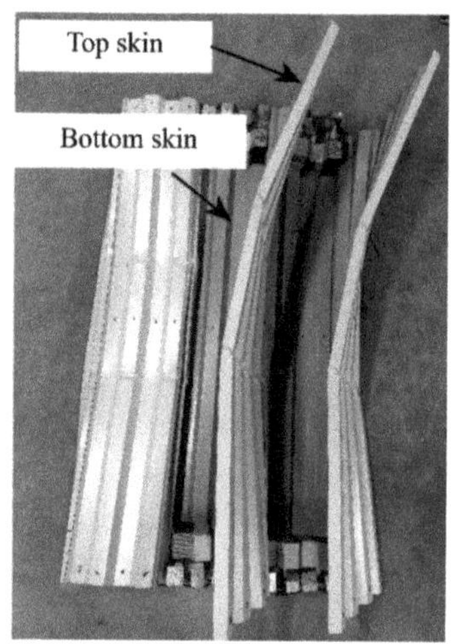

(c)

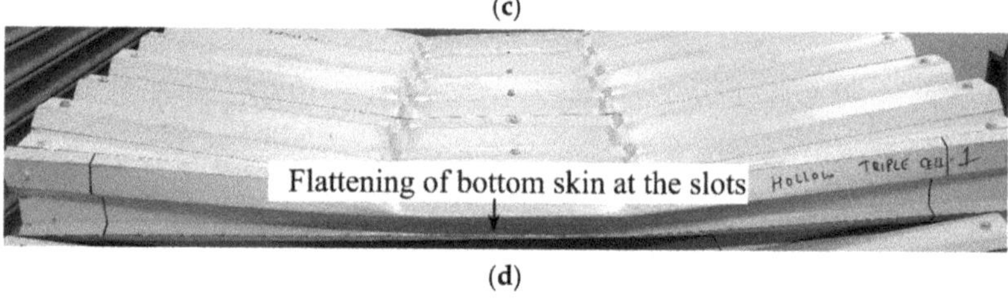

(d)

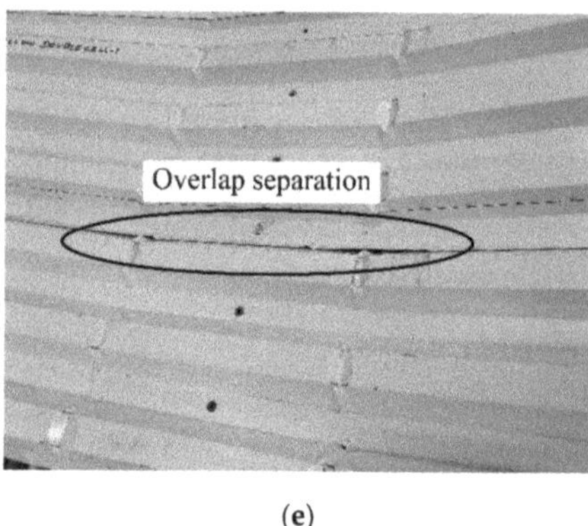

(e)

Figure 9. (**a**) Failed double cell panel. (**b**) Local buckling in triple cell panel. (**c**) Open triple cell sample with the intact bottom skin. (**d**) Flattening of bottom skin at slots. (**e**) Separation at the overlap of double cell-1.

3.2.3. Summary of Failure Mode Results

The failure modes for the four different configurations of multi-cell panels and single cells were analysed. It is found that local buckling was a noticeable failure mode in all the specimens irrespective of the number of cells. The other observed failure type was the slot failure with single-cell containing mid-plates, which might be because of deformation around the mid-plates. However, there was no local buckling in the bottom skin of the single cells without mid-plates which indicates the lack of effective load transfer between the two skins of the panel. In the case of multi-cells, there was no significant separation between the individual cells despite recording local buckling, which can be attributed to the efficient overlay of the sheet panels. It is to be noted that both double and triple cell panels failure modes were similar in all specimens tested. Analysis of failure modes recommends the need for mid-plates or any other infill medium to enhance the efficiency of the proposed panel sections for flooring. It is recommended to fill the voids with some cementitious material so that this system can be further used in structural applications as a flooring system. The infill component will assist efficient use of both top and bottom skins in resisting load and avoiding failure of only top skin.

3.3. Load-Strain Distribution

3.3.1. Single-Cell Panels

The load-strain curves for SC-P and SC panels are shown in Figure 10. The strains were measured at the centre of the panel on the top and bottom skins and underneath a loading point on the bottom skin. In Figure 10a the top of the SC-P panel remains in compression while the bottom of the panel continues in tension due to the action of the mid-plate. The point under the load also remains in tension in the bottom skin. Equal action of top and bottom skins can be observed due to similar strain distribution in both skins with increasing load value.

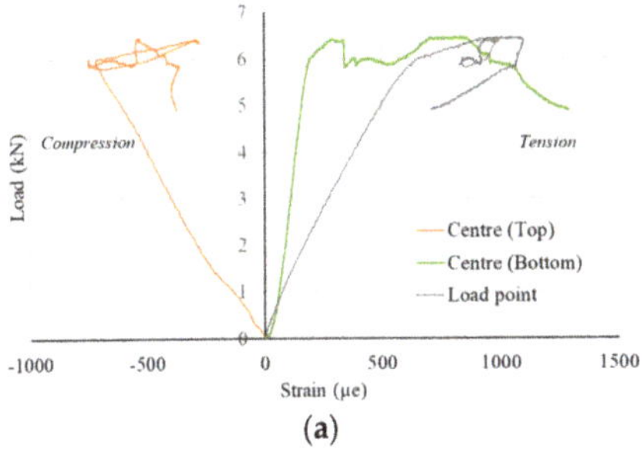

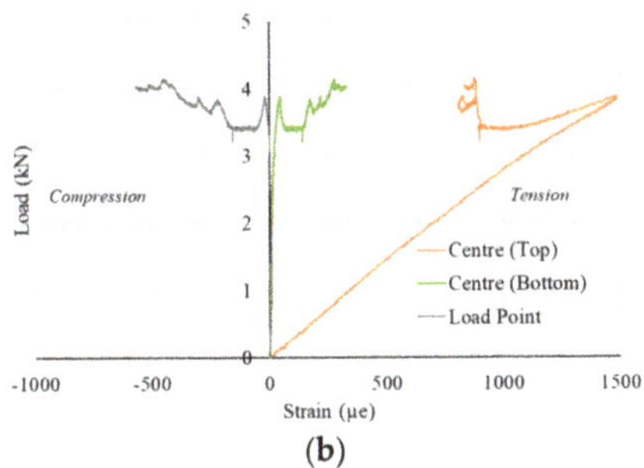

Figure 10. (**a**) Load vs. strain curves of SC-P panel. (**b**) Load vs. strain curves of SC panel.

In contrast, the top skin continues in tension until it touches the bottom skin of the SC panel, after which distortion is seen in the strain (Figure 10b). After this point, the bottom skin experiences minor tensile strain. This behaviour was observed once the top skin touched the bottom skin and can be seen from the green colour plot in Figure 10b. The lack of combined action of the double skin can also be noticed from the strain under the load point, which only began when the ultimate load was achieved. The results indicate that the middle plate in SC-P utilised the entire panel capacity to act as a whole. In contrast, the load-carrying capacity of the SC panels was lower than that of the SC-P panel, where each skin acted independently.

3.3.2. Multi-Cell Panels

Top skin shows a steady rise in tensile strain until it touches the bottom skin, after which a distortion by reduction of strain is noticed in both double and triple cell panels (Figure 11). The bottom skin begins a significant increase in strain only at this point, and the points under the load also begin transferring load at this point. This action is like the SC panels, which is also without a mid-plate. A compression behaviour is recorded under the load point in the triple celled panels because of the extended sheet overlap on the panels.

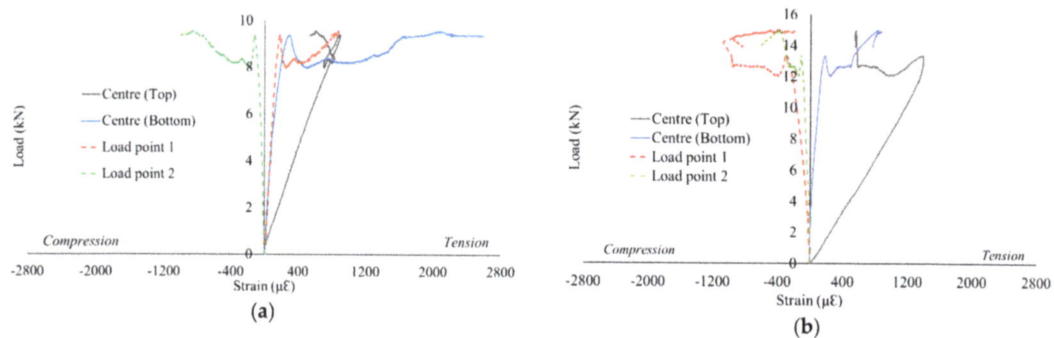

Figure 11. (**a**) Load vs. strain curves of double cell panel. (**b**) Load vs. strain of triple cell panel.

3.3.3. Summary of Load–Strain Behaviour

The load–strain curves were analysed to record the variations in the panel behaviour and identify the critical points of failure. From strain measurements on either side of the panels to monitor the individual behaviour of the top and bottom skins, it is evident that in the panels with mid-plate, the top and bottom skins reported similar behaviour with identical performance in the tension and compression zones. However, in the case of the panels without the mid-plate, the bottom skin started recording significant strains only after the top skin reached its maximum tension capacity. This behaviour indicates the double-skinned action is enhanced and efficient only with mid-plates or any other medium to distribute the loads between the top and bottom skins of the hollow panels.

4. Conclusions

This experimental study confirms the feasibility of adopting the newly proposed innovative flooring systems for temporary structures such as working platforms or formworks. The multi-celled panel systems promise enhanced load carrying capacity due to the strengthening offered by the sheet overlap in the floor design. These hollow panel systems meet the Australian standard specifications for construction load limits showing direct application as a ready to use working platform or for temporary floor components in modular construction. The following points summarise the structural performance of the newly proposed lightweight flooring panels:

(i)   The overlap of sheets improves the stiffness of the panels in bending, thereby increasing the load-carrying capacity.
(ii) Triple cell panels achieved 55% higher loads than double cell panels of the same overall dimensions.
(iii) Panels with mid-plate achieved 76% more load than panels without mid-plate due to simultaneous double skin action.
(iv) The flexural failure, local buckling and failure of slots are prominent types of failure observed in various types of specimen configurations used in this study.

The observed deficiencies in the hollow panels can be rectified by infilling these modular components with sustainable cementitious materials such as ultra-high performance concrete [25], geopolymer concrete [26], etc., for structural flooring applications.

Author Contributions: Conceptualisation, K.H., R.A.-A.; Methodology, K.J., S.R., B.K., M.W., K.H., M.E., N.U., M.R.H., R.A.-A.; Formal Analysis, K.J., S.R.; Investigation, K.J., S.R.; Resources, K.H.; Writing—Original Draft Preparation, K.J., S.R.; Writing—Review & Editing, B.K., R.A.-A.; Supervision, B.K., M.W., K.H., M.E., N.U., M.R.H., R.A.-A.; Project Administration, R.A.-A., B.K.; Funding Acquisition, R.A.-A., M.R.H. All authors have read and agreed to the published version of the manuscript.

Funding: The authors acknowledge the research fund from Deakin University, Victoria, Australia. In addition, the authors acknowledge the in-kind support by industry partners L2U Pty. Ltd. and FormFlow, Australia.

Institutional Review Board Statement: Not applicable.

Informed Consent Statement: Not applicable.

Data Availability Statement: The data presented in this study are available on request from the authors.

Acknowledgments: The authors acknowledge the technical support by the industry partners L2U Pty. Ltd., and FormFlow, Australia. The authors also gratefully acknowledge the Technical Staff—Lube Veljanoski and Michael Shanahan for their support and assistance to complete the experimental investigation at the Structures Laboratory, Deakin University.

Conflicts of Interest: The authors declare no conflict of interest.

References

1. Fiume, F.; Callegaro, N.; Albatici, R. Modular Construction for Emergency Situation: A Design Methodology for the Building Envelope. In *Sustainability and Automation in Smart Constructions*; Springer: Cham, Switzerland, 2021; pp. 131–141. [CrossRef]
2. Gatheeshgar, P.; Poologanathan, K.; Gunalan, S.; Shyha, I.; Sherlock, P.; Rajanayagam, H.; Nagaratnam, B. Development of affordable steel-framed modular buildings for emergency situations (Covid-19). In *Structures*; Elsevier: Amsterdam, The Netherlands, 2021; Volume 31, pp. 862–875. [CrossRef]
3. Zhang, Y.; Lei, Z.; Han, S.; Bouferguene, A.; Al-Hussein, M. Process-oriented framework to improve modular and offsite construction manufacturing performance. *J. Constr. Eng. Manag.* **2020**, *146*, 04020116. [CrossRef]
4. Lin, T.; Lyu, S.; Yang, R.J.; Tivendale, L. Offsite construction in the Australian low-rise residential buildings application levels and procurement options. *Eng. Constr. Archit. Manag.* **2021**. [CrossRef]
5. Dai Pang, S.; Liew, J.R.L.; Dai, Z.; Wang, Y. Prefabricated Prefinished Volumetric Construction Joining Tech-niques Review. *Modul. Offsite Constr. (MOC) Summit Proc.* **2016**. [CrossRef]
6. Hussein, M.; Eltoukhy, A.E.; Karam, A.; Shaban, I.A.; Zayed, T. Modelling in Off-Site Construction Supply Chain Management: A Review and Future Directions for Sustainable Modular Integrated Construction. *J. Clean. Prod.* **2021**, *310*, 127503. [CrossRef]
7. Choi, J.O.; Chen, X.B.; Kim, T.W. Opportunities and challenges of modular methods in dense urban environment. *Int. J. Constr. Manag.* **2019**, *19*, 93–105. [CrossRef]
8. Liew, J.; Chua, Y.; Dai, Z. Steel concrete composite systems for modular construction of high-rise buildings. In *Structures*; Elsevier: Amsterdam, The Netherlands, 2019; Volume 21, pp. 135–149. [CrossRef]
9. Jang, J.; Ahn, S.; Cha, S.H.; Cho, K.; Koo, C.; Kim, T.W. Toward productivity in future construction: Mapping knowledge and finding insights for achieving successful offsite construction projects. *J. Comput. Des. Eng.* **2021**, *8*, 1–14. [CrossRef]
10. Ferdous, W.; Bai, Y.; Ngo, T.D.; Manalo, A.; Mendis, P. New advancements, challenges and opportunities of multi-storey modular buildings—A state-of-the-art review. *Eng. Struct.* **2019**, *183*, 883–893. [CrossRef]
11. Wang, L.; Webster, M.D.; Hajjar, J.F. Behavior of deconstructable steel-concrete shear connection in composite beams. *Struct. Congr. 2015* **2015**, 876–887. [CrossRef]
12. Gunawardena, T. Behaviour of Prefabricated Modular Buildings Subjected to Lateral Loads. Ph.D. Thesis, University of Melbourne, Melbourne, Australia, 2016.
13. Boadi-Danquah, E.; Robertson, B.; Fadden, M.; Sutley, E.J.; Colistra, J. Lightweight modular steel floor system for rapidly constructible and reconfigurable buildings. *Int. J. Comput. Methods Exp. Meas.* **2017**, *5*, 562–573. [CrossRef]
14. Hosseini, M.R.; Chileshe, N.; Rameezdeen, R.; Lehmann, S. Integration of design for reverse logistics and harvesting of information: A research agenda. *Int. J. Logist. Syst. Manag.* **2015**, *20*, 480–515. [CrossRef]
15. Vaz-Serra, P.; Wasim, M.; Egglestone, S. Design for manufacture and assembly: A case study for a prefabricated bathroom wet wall panel. *J. Build. Eng.* **2021**, *44*, 102849. [CrossRef]
16. Razkenari, M.; Fenner, A.; Shojaei, A.; Hakim, H.; Kibert, C. Perceptions of offsite construction in the United States: An investigation of current practices. *J. Build. Eng.* **2020**, *29*, 101138. [CrossRef]
17. Brunesi, E.; Nascimbene, R. Numerical web-shear strength assessment of precast prestressed hollow core slab units. *Eng. Struct.* **2015**, *102*, 13–30. [CrossRef]
18. Hosseini, M.R.; Martek, I.; Zavadskas, E.K.; Aibinu, A.A.; Arashpour, M.; Chileshe, N. Critical evaluation of off-site construction research: A Scientometric analysis. *Autom. Constr.* **2018**, *87*, 235–247. [CrossRef]
19. John, K.; Ashraf, M.; Weiss, M.; Al-Ameri, R. Experimental investigation of novel corrugated steel deck under construction load for composite slim-flooring. *Buildings* **2020**, *10*, 208. [CrossRef]
20. Arrayago, I.; Real, E.; Mirambell, E.; Marimon, F.; Ferrer, M. Experimental study on ferritic stainless steel trapezoidal decks for composite slabs in construction stage. *Thin-Walled Struct.* **2019**, *134*, 255–267. [CrossRef]
21. FormFlow. 2020. Available online: https://formflow.net.au (accessed on 13 December 2021).
22. Madsen, T.; Hansen, K.H. Building System. U.S. Patent 8,539,730 B2, 2013.

23. Lysaght Customflow. 2021. Available online: https://www.lysaght.com/products/customflow (accessed on 13 December 2021).
24. Lysaght. 2021. Available online: https://cdn.dcs.lysaght.com/download/lysaght-build-on-inspiration-design-in-steel (accessed on 13 December 2021).
25. Liao, J.; Yang, K.Y.; Zeng, J.-J.; Quach, W.-M.; Ye, Y.-Y.; Zhang, L. Compressive behavior of FRP-confined ultra-high performance concrete (UHPC) in circular columns. *Eng. Struct.* **2021**, *249*, 113246. [CrossRef]
26. Rahman, S.K.; Al-Ameri, R. A newly developed self-compacting geopolymer concrete under ambient condition. *Constr. Build. Mater.* **2021**, *267*, 121822. [CrossRef]

Article

Insurance Policies for Condition-Based Maintenance Plans of ETICS

Ilídio S. Dias, Ana Silva *, Carlos Oliveira Cruz, Cláudia Ferreira, Inês Flores-Colen and Jorge de Brito

Civil Engineering Research and Innovation for Sustainability (CERIS), Department of Civil Engineering, Architecture and Georesources, IST-University of Lisbon, Av. Rovisco Pais, 1049-001 Lisbon, Portugal; ilidio.dias@tecnico.ulisboa.pt (I.S.D.); oliveira.cruz@tecnico.ulisboa.pt (C.O.C.); claudiaarferreira@tecnico.ulisboa.pt (C.F.); ines.flores.colen@tecnico.ulisboa.pt (I.F.-C.); jb@civil.ist.utl.pt (J.d.B.)
* Correspondence: ana.ferreira.silva@tecnico.ulisboa.pt

Abstract: Currently, insurance companies exclude the buildings' envelope of their policies since they lack reliable information about the risks and degradation models and are unable to estimate the probabilities of intervention and corresponding costs. This study intends to overcome the existing gap, proposing property maintenance insurance policies developed based on condition-based maintenance plans, using stochastic information regarding the degradation process of the buildings' envelope elements in the definition of insurance policies. To perform this work, external thermal insulation composite systems (ETICS) are used as case study, for the definition of an insurance policy. This approach allows reducing the uncertainty associated with the degradation of ETICS even when subject to scheduled maintenance actions. Several insurance policies are analysed, with different insurance premiums, evaluating different risks accepted by the owners when adopting a certain maintenance plan. For owners, the main advantages of acquiring this insurance product are: (i) changing the nature of the risk, transferring the risk to the insurer; and (ii) increasing the asset's equity value, reducing the risk associated with the degradation of ETICS and the uncertainty of maintenance costs over time.

Keywords: insurance policies; maintenance plans; external thermal insulation composite systems (ETICS)

Citation: Dias, I.S.; Silva, A.; Cruz, C.O.; Ferreira, C.; Flores-Colen, I.; de Brito, J. Insurance Policies for Condition-Based Maintenance Plans of ETICS. *Buildings* **2022**, *12*, 707. https://doi.org/10.3390/buildings12060707

Academic Editor: Fabrizio Ascione

Received: 4 April 2022
Accepted: 23 May 2022
Published: 24 May 2022

Publisher's Note: MDPI stays neutral with regard to jurisdictional claims in published maps and institutional affiliations.

1. Introduction

The degradation of buildings and their components is an inevitable process, from the moment they are built and put into use [1]. Maintenance is the most effective way to minimize the building elements' degradation pace as well as prolonging their service life [2]. However, even with a rational and complete maintenance plan, there is always a risk that the building elements deteriorate at a faster pace, compared to what was planned, requiring the anticipation of the necessary maintenance actions, so as not to compromise the durability of these elements. In England [1,3], the buildings' owners are responsible for all the injuries and losses caused by accidents due to anomalies in buildings and their components. The insurance companies exclude from their policies the coverage of any damage caused by the failure of the constructive elements due to lack of maintenance or inadequate maintenance. However, the risks of premature degradation and the uncertainty about maintenance costs can be mitigated by associating insurance policies with the maintenance contract for the building elements.

In this study, exhaustive research was performed, and several studies addressing insurance policies in buildings were found (Figure 1). The existing studies are more concerned with the construction stage [4], with real estate mortgages and warranty schemes [5], with fire and multi-hazard (these insurances are often mandatory in many countries) [6], or with floods, hurricanes, or seismic events [7–10].

Figure 1. Word cloud of the studies found in Web of Science related to insurance policies in buildings.

The insurance market excludes the buildings' envelope of their policies due to the lack of reliable information about the risks of degradation of these components, being unable to predict the probability of failure and the related costs. In this sense, no studies were found in the literature (other than previous studies by the authors [11,12]), nor are real insurance policies associated with the maintenance of façade claddings [13]. This study intends to respond to the lack of insurance related to the risk of anticipated degradation of the building envelope elements. In this sense, an insurance policy associated with a maintenance plan is proposed, to face the risk of needing anticipated maintenance of external thermal insulation composite systems (ETICS), within a 40-year horizon. This investigation begins with a survey of existing insurance products on the market, followed by an analysis of the feasibility of a new insurance product in terms of the potential target market. A methodology for calculating the insurance policy for ETICS is then developed. Maintenance plans are defined through stochastic maintenance models, based on the condition of the element, and using Petri nets, according to the authors' previous work [14,15]. Next, the insurance policies are estimated using the proposed maintenance plans and the scenarios defined for insurance underwriting. Finally, the model is applied to a case study, i.e., the costs are calculated, and a policy is selected for a real building, coated with a current rendering, and with a known maintenance history.

2. Background

2.1. Insurance Policies for External Claddings

The origin of insurance is associated with the concept of reducing the risk of unexpected losses [16,17]. The first foundations of modern insurance emerged in the form of solidarity and compensation to help those affected by misfortune [18]. In the Mesopotamian civilization, there was a system of mutual assistance in case goods transported in caravans did not reach their destination [19]. In the Middle Ages, the concept of "insurance" emerged, with the transfer of the risk of an activity to third parties. In particular, a maritime insurance was conceived, through an association of merchants, in which a contract was established in which everyone agreed to guarantee, for a previously fixed amount, that whoever lost his boat and cargo due to shipwreck would be reimbursed for the loss [20]. In 1666, with the great fire of London, the first insurances for buildings appeared, namely insurance against fire risk [18]. Later, several insurance products emerged to protect customers against unforeseen losses in buildings, mostly aimed at the construction phase of the building and the construction guarantee period [17,21]. In terms of the building use phase, there are generally fire insurance and multi-risk insurance, which protect buildings from accidents of various natures, such as explosion, flood, lightning, seismic phenomena, and civil liability towards neighbours and theft [11,22]. However, it was found that there is practically no insurance to partially protect buildings against damage related to the degradation of the building envelope elements, such as the facade cladding [12].

This study intends to fill this gap by developing an insurance to face the risk of anticipated degradation of the elements of the building envelope, more specifically for external thermal insulation composite systems (ETICS). Therefore, the proposed insurance policies are designed based on the: (i) prediction of the costs associated with claims, considering normal operating conditions and unforeseen situations; and (ii) financial analysis of the risk premium, based on market projections, the expected profit margin, and the industry's average return on investment. Several insurance policies are analysed, with different risk premiums, evaluating different risks accepted by the owners when adopting a maintenance plan. Thus, the proposed insurance policies are adjusted according to the maintenance plan contracted for the building and the age of the coating. For owners, the main advantages of acquiring this type of insurance product are: (i) changing the nature of the risk and its allocation, transferring the risk to the insurer; and (ii) increasing the asset's equity value, reducing the risk associated with the degradation of ETICS and the uncertainty of maintenance costs over time.

2.2. Insurance Target Market

The viability of an insurance operation implies the mass issuance of contracts by the insurance company [23]. This study analysed the potential market for this type of insurance, namely the number of buildings in Portugal with an external coating in ETICS. Based on the latest detailed data made available by the Portuguese Institute of Statistics, the 2011 Census [24], the information on the exterior claddings in buildings was analysed, according to the period of construction and by the main material used in the construction of buildings built until 2011. The data reveal a total of 3,554,389 buildings constructed, and preliminary data from the 2021 census [25] point to an increase of about 1.2% to this value (3,587,669 buildings). The available information reveals that the most used coating on the exterior of buildings in Portugal is traditional renderings and marble, present in about 84% of existing buildings, corresponding to 2,977,132 buildings, built up to 2011. ETICS do not appear individually identified in the statistical data. They are included in the category of other exterior coatings in buildings, present in 0.65% of existing buildings, corresponding to around 23,037 buildings, built up to 2011. This type of coating only began to be used more significantly in Portugal from 2005 onwards, given the new requirements for the thermal behaviour of buildings [25]. Consequently, considering the category of other coatings such as ETICS, and in the absence of more precise data, a market is estimated for around 3288 buildings to be insured, according to the 2011 Census [24] and considering only the period 2006–2011, with an increasing trend for the following decade due to the growing concern about the energy efficiency of buildings.

This insurance product should be more geared towards the maintenance of ETICS and, in this sense, it is also important to consider the repair needs of this type of coating in the building stock. According to data made available by INE on the need for repairs to the exterior walls and frames [24], 64.56% of the existing buildings present a good conservation condition at the level of the exterior walls and did not need to be repaired. On the other hand, 5.93% of the buildings show large and very large repair needs. In this sense, and in the search for an insurance product more focused on the maintenance of the elements of the building envelope, it is possible to correlate the data related to the exterior coatings of the buildings with the repairs needed on the exterior walls and to estimate a possible market for this type of maintenance insurance. Establishing this proportionality for the 3288 buildings with ETICS, there is thus a potential universe of application of this type of insurance of 3270 buildings.

3. Materials and Methods

3.1. Research Method

The research methodology of the present study is illustrated in Figure 2. The first step was a literature review, revealing a lack of studies in this area. This study proposes an innovative insurance policy for the buildings' envelope. As mentioned in the previous sections, the insurance companies are not available to have this type of product due to the lack of reliable data on the degradation of these elements as well as regarding the risk associated with that degradation. Therefore, this research encompasses the following steps: (1) definition of the insurance target market, to analyse the economic viability of this product; (2) creation of a degradation model, based on the assessment of the degradation condition of 378 ETICS, adopting a sigmoidal degradation curve [26,27]; (3) development of a condition-based maintenance model, using Petri nets [14,15]; (4) performance of a risk analysis, based on the condition-based maintenance model, regarding the risk of transition between degradation conditions and the risk related to the anticipation of the maintenance actions and costs, when compared with the maintenance plan under contract; and (5) definition of an insurance policy, considering the maintenance plan contracted and the risks of anticipated maintenance actions.

Figure 2. Research methodology for the definition of an insurance policy for ETICS.

3.2. Methodology for the Definition of an Insurance Policy for ETICS

This study proposes a methodology to combine an insurance policy with a contract for the maintenance of ETICS, throughout their service life, in order to mitigate the risk of anticipated degradation of this element of the building envelope, and to postpone its full replacement. Therefore, in order to guarantee an adequate condition and performance of the elements of the building envelope, maintenance actions must be carried out throughout its service life, according to the condition of the element at a given moment in time. In turn, the design of an insurance policy is directed towards the risk of the need for advance maintenance of ETICS in view of a previously defined maintenance plan. The creation of an insurance policy implies the existence of a risk profile of the object to be insured, so that, later, based on the costs of the actions, the respective annual premium is determined.

In this sense, it is important for the insurer to have information on the insured object and its surrounding environment. In ETICS, these characteristics can be collected through a visual inspection, carried out by the team responsible for the building's maintenance, and with the help of inspection sheets or a computational tool that allows characterising the element's state of degradation at the time of inspection. The evaluation of the overall degradation condition of ETICS is carried out through the calculation of the severity of degradation index (S_w), expressed by Equation (1), according to the service life prediction model initially proposed by Gaspar and de Brito [28]. This numerical index is determined through the ratio between the weighted degraded area and a reference area, equivalent to the entire façade with the highest possible level of degradation [1].

$$S_w = \frac{\sum(A_n \times k_n)}{A \times k} \tag{1}$$

where S_w represents the severity of degradation of ETICS, expressed as a percentage; A_n the area of the element affected by the anomaly n, in m^2; k_n the defect n multiplying factor, in terms of its degradation level, within the $K = \{0, 1, 2, 3, 4\}$ range; A the total façade area, in m^2; and k the multiplying factor corresponding to the highest degradation condition level of the cladded area. This quantitative index (S_w) can be associated to a qualitative scale composed of five degradation conditions: condition A—no visible degradation ($S_w \leq 1\%$); condition B—good ($1\% < S_w \leq 10\%$); condition C—slight degradation ($10\% < S_w \leq 30\%$); condition D—moderate degradation ($30\% < S_w \leq 50\%$); and condition E—generalized degradation ($S_w \geq 50\%$) [26–30]. This system thus makes it possible to assess the degradation condition of ETICS and schedule maintenance actions based on these degradation scales.

In the risk analysis, other important information must also be collected, e.g., the age of the coating, the ETICS' characteristics, and the environmental exposure conditions. In the proposed insurance model, the risk profile is translated into the need to carry out an anticipated action or a deeper maintenance action, which is different from the planned one. Consequently, the loss to be covered by the insurer is related to the anticipation of these actions. The estimation of the costs related to these actions is fundamental in the development of this type of insurance and in the calculation of the respective insurance premium, since it allows estimating the value of future losses. The calculation of the insurance premium thus implies knowing the possible moments in which the indemnities may occur during the life of the element and their costs. Based on the stochastic maintenance models developed for ETICS [14,31], it is possible to access the moments for the need of intervention through the transition probabilities of the degradation condition obtained from these models (this information is described in detail in Section 3). In addition, the value of the cost to carry out these unforeseen actions in the future must be estimated. In other words, financial flows can occur at different times in the future and, therefore, it is necessary to update them to a reference moment. The present value (PV) of the future payment of the claim can be calculated through Equation (2).

$$PV = \frac{C_t}{(1+r)^t} \tag{2}$$

where C_t is the cost, in euros, payable at the end of year t, at discount rate r [32]. Applying the formula to the calculation of the present value of the costs of maintaining the element (ETICS), Equation (3) is used.

$$PV, cost = \sum \left(\frac{C_{t,real\,(insp)}}{\left(1 + r_{real,\,cost}\right)^{t_n}} \right) + \frac{C_{t,real\,(B)}}{\left(1 + r_{real,\,cost}\right)^{t_B}} + \frac{C_{t,real\,(C)}}{\left(1 + r_{real,\,cost}\right)^{t_C}} + \frac{C_{t,real\,(D)}}{\left(1 + r_{real,\,cost}\right)^{t_D}} \tag{3}$$

where $C_{t,real(insp)}$ corresponds to the inspection cost and $C_{t,real(B)}$, $C_{t,real(C)}$, and $C_{t,real(D)}$ correspond to the costs of maintenance actions when the element reaches the degradation conditions B, C, and D, respectively. These actions are further described in the next section. The instants t_B, t_C, and t_D correspond to the years in which the costs associated with maintenance actions occur and the instant t_n to the inspection moments. The real discount rate of the cost, ($r_{real,cost}$), represents the discount of the cost of maintenance actions and relates the nominal discount rate, r_{nom}, and the sector's inflation, i_s, according to Equation (4).

$$r_{real,\,cost} = \frac{1 + r_{nom}}{1 + i_s} - 1 \tag{4}$$

The calculation of the present value of the insurance premium is given by Equation (5). The term $C_{t,premium}$ corresponds to the annual premium payable in €/m^2.

$$PV, premium = \frac{C_{t,premium}}{\left(1 + r_{real,\ premium}\right)^1} + \frac{C_{t,premium}}{\left(1 + r_{real,\ premium}\right)^2} + \ldots + \frac{C_{t,premium}}{\left(1 + r_{real,premium}\right)^{40}} \tag{5}$$

where $r_{real,premium}$ is the real discount rate relative to the premium, which takes into account the nominal discount rate, r_{nom}, the sector's inflation, i_s, and general inflation, i_g (Equation (6)).

$$r_{real,\ premium} = \frac{1 + r_{nom}}{(1 + i_s) \times (1 + i_g)} - 1 \tag{6}$$

The value of the annual premium payable for the insured product, $C_{t,premium}$, is obtained by equalling the net present value of maintenance costs (*PV,cost*) to the present value of the premium (*PV,premium*) (Equations (3) and (5), respectively). The application of this methodology to ETICS and the respective values of each parameter are presented in detail in Section 4.

4. Maintenance Plans for ETICS

4.1. Costs of Maintenance Actions

In this study, four types of maintenance actions are defined to be carried out in ETICS throughout its service life:

- Inspections—€1.03/m^2 (3-year periodicity);
- Cleaning operation (CO)—Condition B (S_w > 1% to 10%)—€26.88/m^2;
- Light maintenance action (LMA)—Condition C (S_w > 10% to 30%)—€58.13/m^2;
- Total replacement (TR)—Condition D (S_w > 30%)—€95.98/m^2.

Cleaning operations are the first action to be carried out on ETICS and intend to eliminate visual and aesthetic anomalies. After an inspection of the façade, which should take place every 3 years, this action is carried out if the coating is in degradation condition B, corresponding to a degradation severity (S_w) between 1% and 10%. This type of intervention generally consists of a water jet cleaning and manual brushing of the surface in order to remove visual anomalies present in the coating, such as surface dirt, stains, and the presence of microorganisms. The costs inherent to this action are presented in Table 1.

Table 1. Cost of cleaning operations in ETICS.

Description	Unity	Yield	Unit Cost	Total Cost
Scaffolding				10.20 €
Hire, assembly, dismantling, transport, and removal of scaffolding. Rental, for 15 calendar days, of standardized tubular scaffolding, multidirectional, up to 10 m maximum working height, for the execution of a façade with 250 m^2	m^2	-	10.20 €	10.20 €
Cleaning operations Cleaning with water jet and manual brushing				16.68 €
Materials				1.70 €
Water	m^3	0.039	1.50 €	0.06 €
Colourless alkaline detergent	L	0.200	-	-
Application of diluted hydrochloric acid solution, with a yield of 0.30 L/m^2, for cleaning efflorescence	L	0.300	-	-
Application of broad-spectrum biocide to eliminate fungi and algae on facades (diluted in water from 20% to 30%)	L	0.119	13.81 €	1.64 €
Contact herbicide for the destruction of herbaceous plants and roots	L	0.005	-	-

Table 1. *Cont.*

Description	Unity	Yield	Unit Cost	Total Cost
Equipment				0.52 €
Low pressure water jet equipment	h	0.097	5.41 €	0.52 €
Hand sprayer for phytosanitary and herbicide treatments	h	0.017	-	-
Labor				14.13 €
1st construction officer	h	0.405	18.48 €	7.19 €
Unskilled construction worker	h	0.447	17.84 €	6.94 €
Complementary direct costs *	%	2	16.35 €	0.33 €
Total cost	m^2			26.88 €

* Equipment and tools associated with the work, which are not contemplated in the unit cost of each item.

The light maintenance action is carried out after an inspection of the facade, when ETICS present an intermediate condition of degradation (condition C), corresponding to a severity of degradation (S_w) between 10% and 30%. This maintenance action includes a cleaning operation (Table 1), repairs, and partial replacement of the coating system. The cost of these actions varies depending on the technique and the area of cladding repaired. The costs for this maintenance action are shown in Table 2. For more details, see [14,31].

Table 2. Cost of light maintenance actions in ETICS.

Description	Unity	Yield	Unit Cost	Total Cost
Scaffolding				10.20 €
Hire, assembly, dismantling, transport, and removal of scaffolding. Rental, for 15 calendar days, of standardized tubular scaffolding, multidirectional, up to 10 m maximum working height, for the execution of a façade with 250 m^2	m^2	-	10.20 €	10.20 €
Cleaning operations Cleaning with water jet and manual brushing				16.68 €
Loss of integrity				10.07 €
Superficial crack repair	%	25	34.45 €	6.55 €
Deep crack repair	%	5	85.78 €	3.43 €
Replacement of deteriorated corners	%	10	1.08 €	0.09 €
Loss of adhesion				10.34 €
Partial replacement of the finishing layer that presents blistering, with possible mechanical reinforcement of the fixing of the plates	%	40	34.45 €	10.34 €
Anomalies in joints	%	40		10.84 €
Partial replacement of the finishing layer and replacement of mechanical fastenings on plates that have cracks in the joints and visualization of joints between plates			36.13 €	10.84 €
Total cost	m^2			58.13 €

The total replacement of the ETICS occurs when the coating reaches its end of service life, corresponding to the degradation condition D (for a severity of degradation equal or higher to 30%), and includes the removal of the existing layer, and the regularization and application of a new coating system [33].

4.2. Maintenance Strategies and Maintenance Action Moments

In order to analyse the impact of maintenance actions on the performance over the lifetime of ETICS, three maintenance strategies are defined [14]:

- MS1: total replacement (TR);
- MS2: light maintenance action (LMA) + total replacement (TR);
- MS3: cleaning operation (CO) + light maintenance action (LMA) + total replacement (TR).

MS1 is defined as the solution most adopted by the owners, where the coating is replaced only when it reaches the end of its service life, with no maintenance during this period. The MS2 is selected to delay or mitigate the coating degradation process in order to maintain adequate building performance and prevent operational interruptions. This strategy intends to demonstrate that small repair actions allow increasing the service life of the element. MS3 adds the cleaning operations to MS2, as it is a maintenance action currently applied to coatings on the building envelope, as it is more economical and less technically demanding. Therefore, based on the stochastic maintenance models previously developed for the three maintenance strategies [21], and considering a probability of occurrence of 50% relative to the transition from the degradation condition of ETICS, the most likely instants (with known probability) of maintenance actions are presented in Tables 3–5. For the moments of maintenance actions in ETICS, a time horizon of 40 years was assumed for the calculation of the insurance value. The time for occurring a maintenance action is given by a probability of 50% for the transition between two degradation conditions, according to the condition-based maintenance model adopted [14,31]. For example, the cleaning operation (CO) occurs when the coating is in condition B, then the time from which such action becomes necessary occurs for a 50% probability of transition between conditions A and B. This risk margin of 50% is considered acceptable by the experts consulted but different risk thresholds (i.e., the probability of ETICS reaching specific degradation values) can adopted. Nevertheless, the risk's reduction is equivalent to an anticipation of the maintenance action's schedule, increasing the maintenance costs and the premium. In a previous study by the authors [11], this risk margin proves to be a realistic compromise between acceptable costs for the insured and the insurer, keeping the coatings in adequate performance conditions.

Table 3. Instants for the total replacement of ETICS considering the first maintenance strategy (MS1), considering a time horizon of 40 years.

Maintenance Action	Condition	50% Probability of Transition between the Degradation Conditions
		Probable Instant of Maintenance Action (Years)
1st TR	C–D	17
2nd TR	C–D	34

Table 4. Instants for the maintenance actions in ETICS considering the second maintenance strategy (MS2), considering a time horizon of 40 years.

Maintenance Action	Condition	50% Probability of Transition between the Degradation Conditions
		Probable Instant of Maintenance Action (Years)
1st LMA	B–C	10
2nd LMA	B–C	18
1st TR	C–D	32

Table 5. Instants for the maintenance actions in ETICS considering the third maintenance strategy (MS3), considering a time horizon of 40 years.

Maintenance Action	Condition	50% Probability of Transition between the Degradation Conditions
		Probable Instant of Maintenance Action (Years)
1st CO	A–B	2
2nd CO	A–B	6
1st LMA	B–C	15
3rd CO	A–B	21
4th CO	A–B	25
2nd LMA	B–C	27

For MS1, two maintenance actions must be carried out on ETICS, in years 17 and 34, when there is a 50% probability of the ETICS reaching its end of service life and requiring a replacement to restore an adequate performance condition. In turn, when applying MS2, the need for a total replacement occurs only once, for a horizon of 40 years. Thus, and according to the probabilities of need of a maintenance action in Table 4, for MS2, two light maintenance actions (LMA) are expected to be performed in years 10 and 18, and a total replacement to occur in year 32. In turn, for MS3, four cleaning operations (CO) are considered in years 2, 6, 21, and 25, and two LI in years 15 and 27 (Table 5).

4.3. Maintenance Plans

Having defined the costs of actions, strategies, and maintenance actions times for a probability of occurrence, the annualized maintenance costs of each strategy for ETICS can be estimated. This calculation is made for a horizon of 40 years, assuming that the maintenance contract is carried out in the zero year of use of the building. In this sense, an update of the costs was carried out, through the present value (PV), in accordance with Equations (3) and (4). To obtain the results, a discount rate of 4%, a sector inflation rate (i_s) of 1.9%, and a general inflation rate (i_g) of 1.8% are considered [33]. In accordance with the strategies defined in the previous section, Tables 6–8 show the values for PV and annualized maintenance costs, considering the different maintenance strategies (MS) and a probability of occurrence of 50%.

Table 6. Annualized cost of maintenance for the first maintenance strategy (MS1), considering a time horizon of 40 years.

Moment of Maintenance Action				Costs		
Condition	Action	Age	Cost of Action	Annualized Cost or Present Value (PV) after 40 Years (€/m^2)	Annualized Cost (€/m^2)
D	TR	17	95.98		
D	TR	34	95.98		
			Σ	124.779	3.286

Table 7. Annualized cost of maintenance for the second maintenance strategy (MS2), considering a time horizon of 40 years.

Moment of Maintenance Action				Costs		
Condition	Action	Age	Action	Annualized Cost or Present Value (PV) after 40 Years (€/m^2)	Annualized Cost (€/m^2)
C	LMA	10	58.13		
C	LMA	18	58.13		
D	TR	32	95.98		
			Σ	146.592	3.860

Table 8. Annualized cost of maintenance for the third maintenance strategy (MS3), considering a time horizon of 40 years.

Moment of Maintenance Action				Costs	
Condition	Action	Age	Action	Annualized Cost or Present Value (*PV*) after 40 Years (€/m^2)	Annualized Cost (€/m^2)
B	CO	2	26.88		
B	CO	6	26.88		
C	LMA	15	58.13		
B	CO	21	26.88		
B	CO	25	26.88		
C	LMA	27	58.13		
			Σ	168.29	4.438

As per the values obtained, the MS1 strategy proves to be the most economical, with an annualized maintenance cost of 3.286 €/m^2 of coated facade, for a 40-year time horizon. This corresponds to a 50% probability of occurrence of the actions at the scheduled time and, thus, to 50% risk of the action being required earlier. More specifically, the condition-based maintenance adopted in this study reveals that, in the universe of the ETICS analysed, half reach the degradation condition D after year 17 and, at that time, a first total replacement of the coating will be necessary. One more action of the same type is required at year 34. These actions correspond to a present value at year 40 of 124.779 €/m^2 and a maintenance cost of 3.286 €/m^2/year. All costs shown are applicable to maintenance strategies carried out after the construction of the building. In the case of MS2, light maintenance actions are also performed in the coating. Thus, for a probability of occurrence of 50%, two light maintenance actions and a total replacement are required during the time horizon, as shown in Table 7. These actions correspond to a present value (*PV*) at year 40 of 146.592 €/m^2 and a cost maintenance fee of 3.860 €/m^2/year. For MS3, the same reasoning is adopted, but cleaning operations are added to the maintenance strategy. Consequently, for a probability of occurrence of 50%, four cleaning operations and two light maintenance actions are necessary during the time horizon. These actions correspond to a 40-year *PV* of 168.519 €/m^2 and a maintenance cost of 4.438 €/m^2/year. Additionally, in the three strategies presented, the costs of periodic inspections to occur every 3 years were included.

Once the costs of the various maintenance strategies are estimated, several maintenance plans are described, their coverage, conditions, advantages, and disadvantages, according to the objectives of the owners or managers of the buildings. The choice of the best solution may involve considering a reduction in the risk of the coating reaching the end of its service life, a reduction in the cost, or the presentation of a better condition of degradation of the ETICS throughout its service life. In other words, the best solution must ensure that the ETICS remains in the most favourable conditions of degradation (A and B) for a longer period. To evaluate this last parameter, the efficiency index, *EI*, can be used, which measures the efficiency of each maintenance strategy, namely the ability to keep the element in a more favourable degradation condition [31]. To evaluate the most advantageous strategy in economic terms, the relationship between the efficiency index and the annualized maintenance cost can be established, and the greater the value of this relationship, the more advantageous the maintenance strategy.

Table 9 presents a summary table where three maintenance plans are presented for ETICS. The annualized maintenance costs were studied for new buildings, i.e., a zero age was considered at the time of contracting the maintenance plan with a duration of 40 years. Of the options proposed, the most economical is the low-cost plan, which costs 3.286 €/m^2/year, and which guarantees the execution of two ETICS replacements at 17 and 34 years. Compared to the maintenance-free option, there is a significant improvement in the efficiency index and the permanence time ($t_{pA,B}$) in the best degradation conditions, from 9.4 years to 24.8 years, in a 40-year model. Additionally, the low-cost plan is also the one with the best ratio between the efficiency index and the annualized maintenance cost (0.21) of all the alternatives. On the other hand, the premium plan presents the highest annualized cost among the options, with a cost of 4.438 €/m^2/year. However, the premium plan shows the best efficiency index of the maintenance strategy, 0.76, i.e., it maintains the ETICS in a better degradation condition throughout its service life. On average, this strategy maintains the coating in condition A and B for another 9 years compared to the low-cost plan. The premium plan provides four cleaning operations and two light maintenance actions on the ETICS up to 40 years of age. On the other hand, it is still the least advantageous solution considering the ratio of efficiency index and annualized cost, with a value of 0.17. The standard plan is the intermediate solution, which, despite being more expensive compared to the first option (low-cost), reduces the risk of the facade cladding needing to be replaced earlier, by carrying out light maintenance actions along its lifespan. In other words, this option allows postponing the replacement by 15 years and keep the ETICS in top condition for another 5 years. In addition, this plan makes it possible to increase the performance of the ETICS compared to the low-cost plan, with an efficiency index of 0.71, through the mitigation of anomalies and the maintenance of a better condition of degradation throughout the service life of the ETICS.

Table 9. Maintenance plans applicable to ETICS.

Maintenance Plan	Maintenance Strategy	Cost of Actions (€/m^2)	End of Service Life (Years)	$tp_{a,B}$ * (Years)	Eficciency Index (EI)	Annualised Mainte-nance Cost (€/m^2)	EI/ Annualised Cost	Plan Application Conditions
Low-cost	MS1$_{prob.50}$ (2 TR)	95.98	17	24.8	0.68	3.286	0.21	Two replacement of the ETICS in years 17 and 34. Applicable up to 40 years of age of the coating. Periodic inspections every 3 years are included.
Standard	MS2$_{prob.50}$ (2 LMA; 1 TR)	58.13; 95.98	32	29.8	0.71	3.860	0.18	Two light interventions carried out, in years 10 and 18, and one TR in year 32. Applicable up to 40 years of age of the coating. Periodic inspections every 3 years are included.
Superior	MS3$_{prob.50}$ (4 CO; 2 LMA)	26.88; 58.13	44	33.8	0.76	4.438	0.17	Four cleaning operations are included, in years 2, 6, 21, and 25, and two light interventions in years 15 and 27. Applicable up to 40 years of age of the coating. Periodic inspections every 3 years are included.
Without maintenance	-	-	17	9.4	0.31	-	-	-

* Simulated for a 40-year horizon.

5. Insurance Policy for ETICS

The degradation process of ETICS with an associated maintenance plan still presents a risk of early degradation. Maintenance contracts cannot address this risk, which is related to the probability that the coating degrades more rapidly and, consequently, that maintenance action will be required earlier than defined in the maintenance plan. There may also be a need for another type of deeper maintenance action in the ETICS at a given moment, and insurance must cover this risk. In this context, several insurance policies are studied to apply to ETICS to protect the owners or managers of buildings from the risk of anticipated degradation of the coatings and a consequent decrease in their service life. In this sense, to define an insurance policy for ETICS, information regarding the probability that the actions will be necessary earlier is essential, in relation to the maintenance plan, and the respective cost of these maintenance actions.

The probabilities of the need for maintenance actions depending on the strategy adopted are presented in Tables 6–8, considering a risk of 50%, as well as the costs of maintenance actions. The conditions applicable to a new policyholder of this type are as follows: (i) any repairs for accidental damage are excluded; (ii) it is not possible to acquire the insurance product if the degradation condition is D or higher, and if previous maintenance actions have not been carried out; (iii) the total replacement occurs when the coating reaches the condition D or higher, as established in the maintenance contract; (iv) light maintenance action occurs when the coating reaches degradation condition C, as defined in the maintenance contract; (v) the cleaning operation takes place when the coating reaches degradation condition B, as established in the maintenance contract; (vi) the insurance policies studied provide coverage for the risk of the ETICS requiring maintenance action earlier than the adopted maintenance plan and are applicable for coatings up to 40 years old; and finally, (vii) the value of the premium to be paid to the insurer must be established according to the age of the coating.

For the development of the insurance policy, several scenarios were studied for the three maintenance plans (MS1, MS2, MS3), with different insurance underwriting times and policy duration, to face the risk associated with the ETICS's age at the time of insurance purchase. As a result, the values for acquiring an insurance product for ETICS at the ages of 5, 10, 20, and 30 years, with policy durations of 5 and 10 years, were calculated. It is also assumed that the maintenance plan is contracted for the same period as the duration of the insurance. The summary of the calculated values can be found in Table 10.

Table 10. Insurance policies for ETICS.

Insurance Start Year	Policy Duration (Years)	Maintenance Plan	VAL Maintenance Plan ($€/m^2$)	Premium with Profit Margin ($€/m^2$/Year)	Premium for a Façade with 77.35 m^2 ($€$/Year)	Condominium Premium ($€$/Year)	Cost of the Mainentenance Plan ($€/m^2$/Year)	Cost of the Maintenance Plan for a Façade with 77.35 m^2 ($€$/Year)
5	5	low-cost	1.9585	1.822	140.90	14.09	0.395	30.53
10	5	low-cost	1.9189	**8.594**	664.77	**66.48**	0.387	29.91
20	5	low-cost	1.9585	1.771	137.01	13.70	0.395	30.53
30	5	low-cost	89.4283	0.444	34.35	3.43	18.023	1394.11
5	10	low-cost	3.6914	3.905	302.07	30.21	0.374	28.96
10	10	low-cost	86.0022	0.218	16.89	1.69	8.722	674.64
20	10	low-cost	3.6914	2.591	200.40	20.04	0.374	28.96
30	10	low-cost	2.7374	0.689	53.27	5.33	0.278	21.47
5	5	standard	54.4519	1.390	107.53	10.75	10.974	848.85
10	5	standard	1.9189	4.380	**338.79**	33.88	0.387	**29.91**
20	5	standard	1.9585	**5.226**	404.21	40.42	0.395	30.53
30	5	standard	93.1119	2.601	201.21	20.12	18.766	1451.53
5	10	standard	56.1848	1.990	153.95	15.39	5.698	440.74
10	10	standard	52.1711	1.039	80.38	8.04	5.291	409.25
20	10	standard	3.6914	4.187	323.88	32.39	0.374	28.96
30	10	standard	94.8805	2.199	170.10	17.01	9.622	744.28
5	5	superior	28.2957	2.130	164.78	16.48	5.703	440.82
10	5	superior	54.4123	1.209	93.49	9.35	10.966	847.69
20	5	superior	52.5693	**5.522**	427.17	42.72	10.595	818.98
30	5	superior	0.9689	5.240	405.35	40.53	0.195	15.10
5	10	superior	77.4319	0.549	42.48	4.25	7.853	607.01
10	10	superior	55.2872	1.523	117.80	11.78	5.607	433.41
20	10	superior	104.6970	1.518	117.44	11.74	10.618	820.76
30	10	superior	2.7374	3.742	289.48	28.95	0.278	21.46

Of the calculations carried out for the moments listed, the most relevant are highlighted, namely the moments of underwriting that lead to a higher value of the insurance premium to be paid by the policyholder, those in which the value of the insurance exceeds the annualized cost of the plan of maintenance, the minimum amount of insurance payable, and the maximum annualized cost of the maintenance plan according to the moment of insurance subscription. The values presented in the table were calculated according to Equations (5) and (6), where a discount rate of 4%, a sector inflation rate (i_s) of 1.9%, and a general inflation rate (i_g) of 1.8% [33] are adopted. A 20% profit margin is also considered in the calculation of the insurance premium, which can change according to the insurer goals concerning the insurance product and the target market.

Overall, the standard plan (MS2) is apparently the most affordable to ensure a policy, as it has the lowest maximum insurance value, at 5.226 $€/m^2$/year, compared to the low-cost plan (MS1) and the superior plan (MS3), in which the maximum amount of insurance payable is 8.594 $€/m^2$/year and 5.522 $€/m^2$/year, respectively. All these values are for policies with a duration of 5 years. In turn, the lowest minimum amount payable is for the low-cost plan, with a value of 0.218 €/m²/year, for insurance underwriting at 10 years of age of the ETICS and a policy duration of 10 years. An analysis of the scenarios studied also reveals a preference for underwriting an insurance policy with a duration of 10 years, since the annual amount of insurance payable is lower, regardless of the maintenance plan implemented in the building's coating. In general, due to the ETICS's subscription age, the adoption of a certain maintenance plan may be more advantageous. However, in average terms, in the analysed scenarios, the low-cost plan is the one with the lowest insurance premium values and the standard plan is the one with higher premiums. The low-cost plan is the one with the highest number of moments in which the insurance premium payable exceeds the annualized cost of the maintenance plan.

In terms of maintenance contracts, the maximum annualized cost of maintenance is reached in the low-cost, with an insurance subscription value of 18.023 €/m²/year, starting at age 30 and lasting 5 years. Similar to the values of insurance premiums, the highest annualized maintenance costs occur when contracting for a period of 5 years. In average terms, the low-cost plan is the one with the lowest annualized maintenance cost values, in line with the values presented in the previous section. In turn, the superior plan has the highest annualized maintenance costs.

The insurance policies in Table 10 must be calculated according to the age of the coating and are applicable for a façade with ages below 40 years, excluding repairs due to accidental damage or vandalism, applying the following specific conditions according to the policy adopted:

- Low-cost policy—Covers the risk of ETICS reaching condition D before the time specified in the plan, with compensation up to the amount of two total replacements;
- Standard policy—Covers the risk of ETICS reaching condition C and D before the time foreseen for the action in the plan, with indemnity in the amount of two light maintenance actions and one total replacement;
- Superior policy—Covers the risk of ETICS reaching condition B, C, and D before the time provided for in the plan, with indemnity in the amount of four cleaning operations and two light maintenance actions.

In order to illustrate the calculation of the insurance premium applied, an example building was used, as shown in Figure 3. The studied facade of the building has a cladded area corresponding to 77.35 m². The building was inspected for the first time in 2016, when it was in a good degradation condition (S_w = 2%). The building construction date is 2011 and consists of ten housing units (i.e., condominium with ten owners).

Figure 3. Case study of an ETICS inspected in 2016.

Considering the date of the building construction, it was assumed that the ETICS is 11 years old. Based on this assumption, an attempt was made to select the most suitable insurance policy and the respective maintenance plan to be adopted by the owners. From the calculations presented in Table 10, the low-cost plan is the one with the lowest premium per joint owner, 1.69 €/year per owner, for a duration of 10 years. However, the selection of the standard plan with a policy at 10 years and duration of 5 years leads to the lowest total annual cost per owner of 36.87 €/year per fraction, i.e., considering the insurance policy and maintenance contract. In case the owners want more favourable conditions over the years (lower degradation conditions), the superior plan is the most suitable solution. The total annual cost per unit owner will be 55.13 €/year, for a duration of 10 years.

In general, this case study reveals that the choice of the best insurance product must be carried out taking into account the age of the facade cladding, the maintenance plan best suited to the owners' requirements, and the intended duration of the policy of insurance. From the analyses carried out, it appears that, regardless of the year of insurance subscription, the options for 10-year policies are in most cases the best option in economic terms.

6. Conclusions

This study defined an insurance policy model applicable to ETICS using stochastic maintenance models, based on the condition of the element. This investigation intends to fill the gap in terms of insurance related to the anticipated degradation of buildings, namely in terms of the surrounding elements. To promote the viability of such insurance, an application universe of around 3270 buildings in Portugal was estimated.

This study proposes the definition of real insurance policies, considering the risk associated with the failure of ETICS (and can be adaptable to other building envelope elements), as well as the most probable failure time and the costs associated with that failure. These insurance policies are deemed to significantly enhance the market for outsourcing maintenance, which has been limited due to lack of knowledge in terms of risk and costs and also due to the absence of risk mitigation strategies (insurances) for providers of maintenance services.

The design of this insurance assumes its association with a maintenance contract, since the policy intends to bear the risk of a maintenance action being necessary sooner. In this sense, three maintenance plans were developed for a time horizon of 40 years. The selection of the best plan varies according to the criteria of the owner or manager of the building, but the solution that presents the best cost-benefit ratio (annualized maintenance cost/efficiency index) is the low-cost plan with an index of 0.21.

The study of the different insurance policies applied to the ETICS, depending on the three maintenance plans, reveals that the standard plan is the most economical to ensure a policy, considering only the maximum amount of insurance payable. The minimum amount payable is in the low-cost plan, with a value of 0.218 €/m²/year, for insurance underwriting at 10 years of age of the ETICS and a policy duration of 10 years. An analysis in average terms for all the underwriting moments studied shows that the low-cost plan is the one with the lowest insurance premium values. Regardless of the maintenance plan implemented in the building, subscribing a 10-year policy is almost always more advantageous as it leads to lower insurance premiums.

Through application to the case study, for a 11-year-old ETICS, the low-cost plan is the most economical solution, with an insurance premium per owner of 1.69 €/year, for a 10-year subscription. However, the selection of the standard plan with a policy at 10 years and duration of 5 years leads to the lowest total annual cost per owner of 36.87 €/year per fraction (including annualized maintenance costs and insurance policy cost). In summary, the selection of the best insurance product must consider the age of the coating, the maintenance plan chosen by the owner, and the intended duration of the insurance policy, taking into account that a 10-year subscription is, generally, the most economical option.

This study and the insurance policies proposed present some advantages, namely: (i) changing the nature of the risk and transferring increased risks to the insurance company; and (ii) increasing the patrimonial value of the asset, through the adoption of adequate and timely maintenance actions, reducing the risk of failure of the façades and reducing the cost of urgent and unpredictable maintenance actions. The acquisition of this product by a residential condominium provides a way of share the costs of the maintenance contract for the buildings' façades, sharing the risks by the households, resulting in a lower price for the insurance premium. The major limitation of this insurance product is the lack of funds by the owners to have a maintenance plan and insurance to cover the risk of anticipating the actions. However, not purchasing this type of product entails higher costs in the long term, with the deterioration of the facades (with an increase in Municipal taxes) and the need for costlier unexpected and urgent maintenance actions. Another barrier for the application of this product can be the long-term profits of the insurer. Nevertheless, this study provides an accurate tool to estimate the risk margins, which can lead to more competitive insurance premium leading to larger number of clients in the portfolio, thus reducing the risks assumed by the insurance company.

In future studies, the insurance policies for condition-based maintenance plans will consider the combination of various elements present on the facade, carrying out maintenance actions opportunistically, and reducing the fixed costs of actions (e.g., scaffolding).

Author Contributions: Conceptualization, I.S.D., A.S. and C.O.C.; formal analysis, I.S.D. and A.S.; writing—original draft preparation, I.S.D. and A.S.; writing—review and editing, I.S.D., A.S., C.O.C., C.F., I.F.-C. and J.d.B. All authors have read and agreed to the published version of the manuscript.

Funding: This research was partially funded by the FCT (Portuguese Foundation for Science and Technology) through the project BestMaintenance-LowerRisks (PTDC/ECI-CON/29286/2017) and the PhD programs UI/BD/151150/2021.

Institutional Review Board Statement: Not applicable.

Informed Consent Statement: Not applicable.

Data Availability Statement: The data presented in this study are available on request from the corresponding author.

Acknowledgments: The authors gratefully acknowledge the support of the CERIS Research Centre (Instituto Superior Técnico-University of Lisbon).

Conflicts of Interest: The authors declare no conflict of interest.

References

1. Silva, A.; de Brito, J.; Gaspar, P.L. *Methodologies for Service Life Prediction of Buildings: With a Focus on Façade Claddings*, 1st ed.; Springer: Geneva, Switzerland, 2016.
2. de Brito, J.; Pereira, C.; Silvestre, J.D.; Flores-Colen, I. Expert Knowledge-Based Inspection Systems. In *Inspection, Diagnosis and Repair of the Building Envelope*; Springer: Cham, Switzerland, 2020.
3. Blong, R. Residential building damage and natural perils: Australian examples and issues. *Build. Res. Inf.* **2004**, *32*, 379–390. [CrossRef]
4. Al-Kasasbeh, M.; Abudayyeh, O.; Olimat, H.; Liu, H.X.; Al Mamlook, R.; Alfoul, B.A. A Robust Construction Safety Performance Evaluation Framework for Workers' Compensation Insurance: A Proposed Alternative to EMR. *Buildings* **2021**, *11*, 434. [CrossRef]
5. Royal, S.; Lehoux, N.; Blanchet, P. Comparative case study research: An international analysis of nine home warranty schemes. *Int. J. Build. Pathol. Adapt.* **2021**. *ahead-of-print*. [CrossRef]
6. Landry, C.E.; Anderson, S.; Krasovskaia, E.; Turner, D. Willingness to Pay for Multi-peril Hazard Insurance. *Land Econ.* **2021**, *97*, 797–818. [CrossRef]
7. Khanduri, A.C.; Morrow, G.C. Vulnerability of buildings to windstorms and insurance loss estimation. *J. Wind. Eng. Ind. Aerodyn.* **2003**, *91*, 455–467. [CrossRef]
8. Lamond, J.; Penning-Rowsell, E. The robustness of flood insurance regimes given changing risk resulting from climate change. *Clim. Risk Manag.* **2014**, *2*, 1–10. [CrossRef]
9. Bonstrom, H.; Corotis, R.B. Building portfolio seismic loss assessment using the First-Order Reliability Method. *Struct. Saf.* **2015**, *52*, 113–120. [CrossRef]

10. Filippova, O.; Xiao, Y.; Rehm, M.; Ingham, J. Economic effects of regulating the seismic strengthening of older buildings. *Build. Res. Inf.* **2018**, *46*, 711–724. [CrossRef]
11. Macedo, M.; de Brito, J.; Silva ACruz, C.O. Design of an Insurance Policy Model Applied to Natural Stone Facade Claddings. *Buildings* **2019**, *9*, 111. [CrossRef]
12. Macedo, M.; de Brito, J.; Cruz, C.O.; Silva, A. Methodological proposal for the development of insurance policies for building components. *CivilEng* **2020**, *1*, 1–9. [CrossRef]
13. Forster, A.M.; Kayan, B. Maintenance for historic buildings: A current perspective. *Struct. Surv.* **2009**, *27*, 210–229. [CrossRef]
14. Ferreira, C.; Dias, I.S.; Silva, A.; de Brito, J.; Flores-Colen, I. Criteria for selection of cladding systems based on their maintainability. *J. Build. Eng.* **2021**, *39*, 102260. [CrossRef]
15. Ferreira, C.; Neves, L.C.; Silva, A.; de Brito, J. Stochastic Petri net-based modelling of the durability of renderings. *Autom. Constr.* **2018**, *87*, 96–105. [CrossRef]
16. Mossin, J. Aspects of rational insurance purchasing. *J. Political Econ.* **1968**, *76*, 553–568. [CrossRef]
17. Cruz, C.O.; Branco, F. Reconstruction Cost Model for Housing Insurance. *J. Leg. Aff. Disput. Resolut. Eng. Constr.* **2020**, *12*, 05020007. [CrossRef]
18. Haueter, N.V. A history of insurance. In *Swiss Re Corporate History*; Swiss Re: Zurich, Switzerland, 2013.
19. McLaughlin, R. *Rome and the Distant East: Trade Routes to the Ancient Lands of Arabia, India and China*; Bloomsbury Publishing: London, UK, 2010.
20. Noussia, K. *The Principle of Indemnity in Marine Insurance Contracts: A Comparative Approach*; Springer Science & Business Media: Berlin/Heidelberg, Germany, 2007.
21. El-Adaway, I.H.; Kandil, A.A. Construction risks: Single versus portfolio insurance. *J. Manag. Eng.* **2010**, *26*, 2–8. [CrossRef]
22. Botzen, W.J.W.; Van den Bergh, J.C.J.M.; Bouwer, L.M. Climate change and increased risk for the insurance sector: A global perspective and an assessment for the Netherlands. *Nat. Hazards* **2010**, *52*, 577–598. [CrossRef]
23. Dorfman, M.S. *Introduction to Risk Management and Insurance*, 6th ed.; Prentice Hall International, Inc.: Hoboken, NJ, USA, 1998.
24. National Institute of Statistics. Portugal, 2011. Census 2011, Definitive Results. Available online: https://censos.ine.pt/xportal/xmain?xpid=CENSOS&xpgid=censos2011_apresentacao (accessed on 1 March 2021).
25. National Institute of Statistics. Portugal, 2021. Census 2021, Preliminary Results. Available online: https://www.ine.pt/scripts/db_censos_2021.html (accessed on 1 March 2021).
26. Amaro, B.; Saraiva, D.; de Brito, J.; Flores-Colen, I. Inspection and diagnosis system of ETICS on walls. *Constr. Build. Mater.* **2013**, *47*, 1257–1267. [CrossRef]
27. Tavares, J.; Silva, A.; de Brito, J. Computational models applied to the service life prediction of ETICS. *J. Build. Eng.* **2020**, *27*, 100944. [CrossRef]
28. Gaspar, P.L.; de Brito, J. Quantifying environmental effects on cement-rendered facades: A comparison between different degradation indicators. *Build. Environ.* **2008**, *43*, 1818–1828. [CrossRef]
29. Ximenes, S.; De Brito, J.; Gaspar, P.L.; Silva, A. Modelling the degradation and service life of ETICS in external walls. *Mater. Struct.* **2015**, *48*, 2235–2249. [CrossRef]
30. Marques, C.; de Brito, J.; Silva, A. Application of the factor method to the service life prediction of ETICS. *Int. J. Strateg. Prop. Manag.* **2018**, *22*, 204–222. [CrossRef]
31. Ferreira, C.; Silva, A.; de Brito, J.; Dias, I.S.; Flores-Colen, I. Condition-Based Maintenance Strategies to Enhance the Durability of ETICS. *Sustainability* **2021**, *13*, 6677. [CrossRef]
32. Brealey, R.; Myers, S.; Allen, F. *Principles of Corporate Finance*, 10th ed.; McGraw Hill: New York, NY, USA, 2010.
33. National Institute of Statistics. Portugal, 2021. Consumer Price Index. Available online: https://www.ine.pt/xportal/xmain?xpgid=ine_tema&xpid=INE&tema_cod=1314&xlang=pt (accessed on 1 March 2021).

Article

Integrated Building Maintenance and Safety Framework: Educational and Public Facilities Case Study

Kun-Chi Wang [1], Reut Almassy [2], Hsi-Hsien Wei [3] and Igal M. Shohet [1,4,*]

[1] Department of Civil and Construction Engineering, Chaoyang University of Technology, Taichung 413, Taiwan; wkc@cyut.edu.tw
[2] Safety Management and Engineering Unit, Department of Civil and Environmental Engineering, Faculty of Engineering Sciences, Ben-Gurion University of the Negev, Beer-Sheva 8410501, Israel; reut@br7.org.il
[3] Department of Building and Real Estate, Hong Kong Polytechnic University, Hong Kong, China; hhwei@polyu.edu.hk
[4] Department of Civil and Environmental Engineering, Faculty of Engineering Sciences, Ben-Gurion University of the Negev, Beer-Sheva 8410501, Israel
* Correspondence: igals@bgu.ac.il

Abstract: The facility safety is a highly important issue in educational institutions and public facilities, where the safety and health of the occupants (students, educational and public service staff) is a high-order priority. The research hypothesizes that a synergy exists between the maintenance and safety of public facilities. Analytical–empirical research methods were aimed at the development of an integrated maintenance–safety framework. The framework was validated through inferential statistics and a case study. A correlation (R^2) of 0.74 between the level of maintenance and the safety level of 24 educational facilities was found using the Pearson correlation coefficient. The levels of maintenance and of safety observed were marginal. An innovative Integrated Safety–maintenance performance framework was developed for synergetic safety–maintenance monitoring, control and management. The framework proposes a cycle loop of safety–maintenance–performance audits of facilities as a key tool for advanced maintenance and safety management in public facilities. The framework was validated in a case study of public facility. The time history of maintenance performance and safety shows a high level of fitness ($R^2 = 0.8865$, p-Value < 0.05). These research findings stress that integrated safety and maintenance should be implemented as a unified procedure to enhance the advanced maintenance performance and safety climate in public facilities management.

Keywords: facilities management; maintenance; performance; risk; safety

Citation: Wang, K.-C.; Almassy, R.; Wei, H.-H.; Shohet, I.M. Integrated Building Maintenance and Safety Framework: Educational and Public Facilities Case Study. *Buildings* **2022**, *12*, 770. https://doi.org/10.3390/buildings12060770

Academic Editor: Ana Silva

Received: 26 April 2022
Accepted: 30 May 2022
Published: 5 June 2022

Publisher's Note: MDPI stays neutral with regard to jurisdictional claims in published maps and institutional affiliations.

1. Introduction

Facility safety is highly important in educational institutions and public facilities in particular, where the safety and health of the occupants (students, educational and public service staff) is a high order of priority [1,2]. The safety of users of buildings and facilities is a highly important facility management issue. During the construction of the building, design methods and materials will be reviewed for different types of buildings and facilities to ensure the sustainability of the facilities, as well as the safety of users of the buildings and facilities [3–5]. In addition to the consideration of the construction process, the proactive management of maintenance and performance, as well as safety assessment of the facility during the maintenance phase, are required in order to effectively improve the safety mode of the building [6]. However, for the safety assessment of building facilities, there is currently no definite assessment methodology or indicator, nor a set of valid assessment models, for the implementation of these procedures [7]. To this end, this study summarizes the development of an integrated risk-informed methodology aimed at the integration between the maintenance and safety of facilities as a performance management tool for the fostering of the maintenance and safety of facilities, taking advantage of the synergy between these disciplines. The research develops a model for evaluating and managing building safety and maintenance through this integrated methodology of maintenance and safety indicators.

The development of methods for safety assessments in educational institutions is a ubiquitous research topic, Al-Hemoud and Al-Asfoor [2] developed the Behavioral-Based Safety (BBS) that introduces safety using a behavioral safety approach. This theory follows Heinrich's [8] theory of the human factor in accident causation and stresses the training of individuals, safety control and inspection. DePasquale and Geller [9] examined the success of BBS models from the workers' perspective and found five predictive parameters of the success of the method: (a) effective training, (b) senior management trust, (c) effective safety control, (d) training of workers and (e) workers' commitment. Elementary and high school educational institution occupants are heterogenic groups with diverse patterns of safety behavior assimilation and, therefore, are more vulnerable to safety hazards [10–12]. Sabatino et al. [13] stressed that a methodology for the optimal maintenance of buildings must integrate into its considerations the consequences of a system failure or poor performance. Lai and Yik [14] studied the parameters of the optimal maintenance setup that will provide safety, efficiency and high user satisfaction within the building's occupants. Through a survey of 279 questionnaires in residential buildings in Hong Kong, the researchers analyzed five parameters of maintenance services: safety, cleanliness, break down and routine maintenance, aesthetics and overall satisfaction. This research did not consider the building performance as a key parameter of an optimal maintenance plan. Olanrewaju et al. [15] studied resource allocation priority settings in university facilities in Malaysia. The researchers found out that breakdown maintenance of visible components such as elevators, power supply and roofing failures are a high-order priority and should receive higher precedence in the allocation of routine maintenance resources. This review emphasizes the ultimate role of facility maintenance in ensuring the occupants' safety and satisfaction, as the latter is highly important and crucial in educational institutions and public facilities.

The research main objective is to investigate the correlation between maintenance and safety performance in educational and public facilities as a basis for the development of integrated maintenance and safety framework. Six research subgoals are defined as follows:

a. Definition of the risk factors, according to the facility components, performance and users;
b. Development of a methodology for the assessment of the safety risks in educational and public facilities;
c. Assessment of the maintenance and safety performances in educational and public facilities;
d. Development of an Integrated Safety–Maintenance management framework;
e. Implementation of the framework and validation of the research hypotheses;
f. Recommendations for further research and concluding remarks.

Subsequent from the research hypotheses, findings and conclusions, a new framework for integrated risk-informed maintenance and safety surveys and performance assessments is introduced as a part of the facility management. The framework is designated to take advantage of the synergy between safety and maintenance so as to enhance the maintenance, performance and safety of public facilities. This proposed framework will stimulate lower probabilities of safety hazards in the performances of public facilities; improve the sustainable performance of the facilities on a continuous basis and can fit into other types of facilities such as energy, healthcare, residential buildings and off-shore facilities. While previous research on the topic of safety and maintenance focused on the role of the human factor, in the synergy between maintenance and safety, e.g., Hobbs and Williamson [16] in the Aeronautics industry and Farrington-Darby et al. [17]. In the railway maintenance industry, this research develops a systemic and inherent synergy and introduces an integrated performance control of safety and maintenance management in public and educational facilities.

2. Literature Review

The literature review develops a research theme based on the definitions of the core topics of the research, review of the risk factors maintenance and safety indicators in facilities.

The core variables of this study, encompassing safety, maintenance, risk, acceptable risk and building performance, were defined on the basis of the literature, as follows in Table 1:

Table 1. Cross-comparative definitions of the core variables of the research.

Parameter	Definition	Cross-Comparative Definition
Safety	All preventive measures aimed at ensuring user well-being and facility-designated use.	Hollnagel [18] indicated that safety is accomplished through the prevention of hazards, unsafe events and the consequences of structural failures.
Maintenance	Activities carried out to maintain the durability and performance of the building in accordance with its designated use.	Three aspects of key facility maintenance: (1) actual service condition of the system, (2) failures affecting the service condition provided by the system and (3) actual preventive activities carried out to maintain acceptable designated service condition [19].
Risk	The probability and severity of an undesired event.	Aven [20] stated that the capability to express risk quantitatively and compare alternative mitigation strategies is at the heart of risk management.
Building performance	The concept in which the functions of buildings are defined by their outcomes rather than by prescription. Performance is determined by a building's success in achieving user satisfaction and meeting designated performance criteria.	The evaluation of the overall service condition of the building or of the building portfolio, according to the performance criteria of its components and systems defined by its designated use and current standards [19].
Acceptable risk	The level of risk deemed acceptable for an organization.	Hollnagel [21] stated that risk and safety are conceptually and practically linked—with higher safety meaning lower risk—and that the most effective way to achieve safety is through proactive mitigation actions.

The definitions stem from the performance concept and are supported by cross-referencing from the literature.

2.1. Risk Factors

Risk assessment and management is a key driver of building performance and maintenance management [15,17]. Using the fatigue design factor, Straub and Faber [22] developed a risk-based inspection strategy for planning an optimal inspection regime for structural components. Khalil et al. [23] developed integrated indicators of a building performance and user risk through the analytical hierarchy process method. Their methodology yielded a building performance risk rating tool (BPRT). They identified five critical risk indicators: (1) structural stability, (2) fire protection services, (3) building-related pathology, (4) emergency exits and (5) electrical services. The BPRT was implemented in the maintenance of higher education buildings in Malaysia. Khan et al. [24] introduced a risk-based maintenance methodology composed of three modules: (1) risk estimation, (2) risk evaluation and (3) maintenance planning. They used a heating, ventilation and air-conditioning (HVAC) system as a case study of risk-based maintenance to validate their methodology. The case study contributed to the successful reduction of risk in an HVAC system from an unacceptable level to an acceptable level.

Backlund and Hannu [25] compared three strategies of systems risk assessment to identify a maintenance strategy that minimizes maintenance costs as a part of the deregulation of the Swedish hydro power plants sector. The researchers identified several key maintenance-related tasks for accurately assessing risk: (1) exposure time, (2) frequency estimation, (3) consequence estimation, (4) qualitative and quantitative methods for reliability and frequency estimation, (5) teamwork, (6) performance analysis and (7) uncertainty and sensitivity analysis. Khan and Haddara [26] developed effective risk-based maintenance for HVAC systems by using a scenario analysis, probabilistic failure analysis, risk estimation and evaluation and maintenance planning.

De Silva et al. [27] identified 10 maintenance-related risk factors for buildings: (1) architecture and design risks; (2) structural and detailing risks; (3) service integration risks; (4) accessibility risks; (5) maintenance requirement risks; (6) material and spare parts risks; (7) constructability risks; (8) maintenance process quality risks; (9) characteristics, environment and exposure risks and (10) user requirement and change-related risks. De Silva et al. [27] emphasized the integration of various characteristics of the exposure of the building environment to hazards. Addressing such exposure requires the integration of risk management, safety and maintenance. Ruparathna et al. [28] developed a risk-based scenario planning approach that incorporates predicted changes in technology, costs and organizational strategies for the selection of maintenance strategies. Their methodology minimizes financial risk. Their study also stressed the integration between risk and maintenance planning.

Chiu et al. [29] used the concept of the reliability of a series system to develop a theory of the deterioration risk of Reinforced Concrete structures exposed to chloride attacks. They determined an optimal maintenance plan using probabilistic effect assessment models that considered the effects of maintenance strategies on the failure and spalling probability of a structure and could be used to estimate life cycle costs and performances. Their study emphasized the need to integrate safety and maintenance in a comprehensive framework.

2.2. Maintenance

Despite being part of the last stage of the facilities life cycle, the maintenance and management of buildings should be considered throughout the design stage. Such considerations should include the intensity of the facility use, facility attributes, facility maintenance methods, facility maintenance costs, facility vulnerabilities, the total number of facilities and the facility maintenance period [30–32]. That is, the impact of the building material configuration on the building facilities (including durability and replaceability) must be considered in the building design. If these factors are considered in detail in the design stage, a building can be maintained and managed at low cost [33].

The quality of buildings and their facilities is often poor because of a lack of proper building maintenance and management, even with high maintenance costs [34–36]. In addition, once passive management is adopted, various types of building defects, including structural or nonstructural cracking and concrete fragments spalling, can cause severe damage. Therefore, many studies have proposed building maintenance and management methods that fall into one of three categories: corrosive, preventive or predictive maintenance. Among these, corrosive maintenance is passive and involves repairing equipment after it is damaged, whereas the other two types of maintenance are active [30]. An optimization framework was developed to consider the interdependencies among different units of HVAC equipment and interdependencies among maintenance, operation and equipment durability to deduce predictive and preventive maintenance.

Saqib, Faruoqui and Lodi [37] identified 77 factors and classified them into seven categories, including project management–related factors, procurement-related factors, client-related factors and business-related factors. They revealed that the 10 CSFs (Critical Success Factors) of a project are decision-making effectiveness, project manager experience, contractor cash flow, contractor experience, owner or owner representative decision timeliness, site management, supervision, planning effort, project management experience and client decision-making ability. They determined that the five CSF groups that most influence project success are contractor-related factors, project manager–related factors, procurement-related factors, design team–related factors and project management-related factors. Au Yong et al. [38,39] validated the importance and role of safety for setting the priority of maintenance activities.

2.3. Safety Indicators

Safety indicators are the core and key tool for the control and management of safety in the facilities life cycle. Ho et al. [40] developed a system composed of a building health and hygiene index (BHHI) and a building safety and conditions index (BSCI) for the improvement of tall residential building performance, hygiene and safety in Hong Kong. They surveyed 140 tall residential buildings and used their survey results to develop an assessment method based on a hierarchy of indicators: (1) architectural design, (2) building services design, (3) the surrounding environment, (4) operations and maintenance and (5) management. They concluded that management systems account for 82% of the variance in the BHHI and 45% of the variance in the BSCI. Their results emphasized the importance of building management systems for the safety and health of building users, particularly in densely populated areas. Hopkins [41] stressed that process safety indicators should measure the effectiveness of the risk control system and that these indicators must be included in an incentive payment scheme. Nevertheless, Hopkins emphasized that careful attention is required to avoid managing the management process rather than actual safety. Following Hopkins, Harms-Ringdahl [42] determined that the core safety indicators can be static or dynamic and should reflect the organizational safety process, as well as the outcomes. Ryczyński et al. [43] proposed that, to strengthen structural safety and durability, special attention must be paid to the causes of cracks, as well as to planning and conducting appropriate repairs. In addition, many building safety accidents are caused by faulty designs. Therefore, risk assessment in the design stage is crucial in risk management [44].

This review of the concepts of risk factors, maintenance and safety indicators reveals a gap in the integration between risk, maintenance and safety for synergetic and integrated maintenance and the safety management of facilities and indicates the research gap. The latter derived the principal purpose of the research: the development of an integrated maintenance–safety framework, as well as the subgoals, to examine the correlation between safety and maintenance in educational and public facilities.

The research hypotheses were derived from the research gap and literature as follows:

a. A correlation exists between the maintenance performance of the educational facility and its safety, i.e., the higher the maintenance performance of the facility, the higher the safety of the facility is.

b. There exists a correlation between the density of occupants (occupancy) and the performance and safety of educational and public facilities.

3. Integrated Risk-Informed Safety–Maintenance Framework

An integrated risk-informed safety–maintenance framework was developed within the scope of the research. The framework is composed of the performance model (monitoring the adequacy of the building to its designated performance), maintenance model (monitoring preventive and breakdown maintenance) through the MI and safety model (monitoring and control of the safety of the building's systems) through the BRI. The framework is implemented using three phases: safety monitoring and control using the BRI, building performance monitoring and building maintenance monitoring using the MI. Synthesis between the safety, maintenance and performance indicators yields control loop remarks and outputs for the implementation of corrective safety maintenance activities and corrective condition-based maintenance that stimulates the synergy between safety and maintenance performance over the life cycle of the building (Figure 1). The framework is implemented with a structured object class demonstrated in Figure 2; it is composed of structural stability, structural integrity, preventive maintenance, breakdown maintenance and a performance evaluation of the components (Figure 3, referring to the exterior envelope). Similar schemes are implemented for the interior finishing; electrical system; water and sewage system; HVAC (Heating, Ventilation and Air Conditioning); fire extinguishing, elevators; Information and Communication Systems (ICT) and peripheral infrastructures.

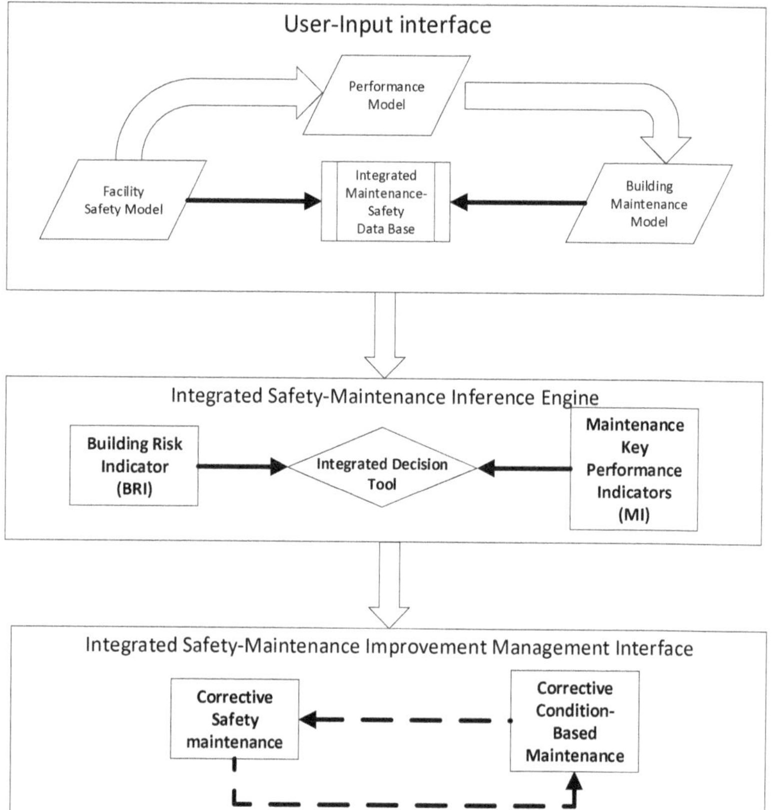

Figure 1. Integrate the risk-informed safety–maintenance implementation framework.

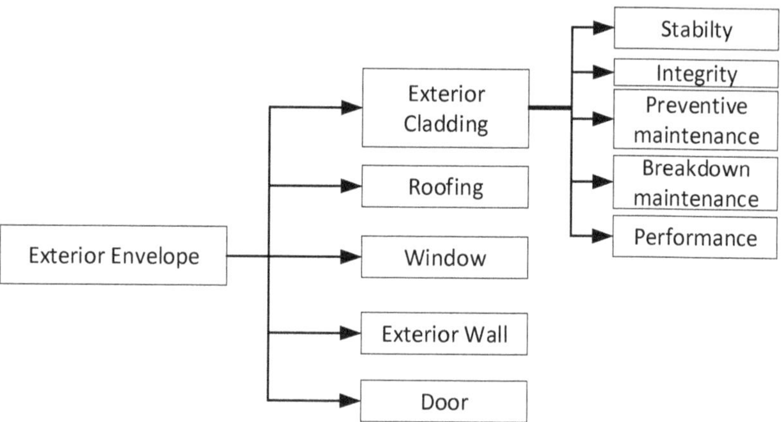

Figure 2. Structure of object classes for parameters of the integrated safety–maintenance in the user-Interface.

Figure 3. Research phases and methodology.

The proposed framework develops a novel risk-informed methodology for the integrated safety–maintenance management of public and educational facilities. The framework builds upon an integrated tool for combined safety–maintenance monitoring of facilities. The framework takes advantage of the synergy between maintenance and safety and leverages the synergy to introduce enhanced high-performance, safe and cost-effective maintenance of facilities.

4. Research Method

The research employed analytical–empirical methods aimed at the development of an integrated maintenance–safety framework and validation of the framework through inferential statistics and a case study. The research followed the definitions of ISO/DIS 45001 [45] and the guidelines of ISO 14001 and OSH 2001: Guidelines on Occupational Safety and Health Management Systems. The research phases and method are described in Figure 3. It followed five steps [19,46,47]: (1) data gathering methodology for safety and maintenance assessment and control in educational and public facilities, (2) field survey of maintenance and safety in educational facilities, (3) inferential statistical analysis and the development of an integrated risk-informed safety–maintenance framework, (4) implementation and validation of the framework in a public facility case study and assessment of the research hypothesis using regression analysis and F-statistics and (5) conclusion and recommendations for further research. The framework is a theoretical and practical framework, and the theory is validated In a field survey in Section 3 and implemented for practical validation in Section 4.

The research employed four tools of data gathering and analysis:

a. Field survey of 24 educational institutions facilities, with at least 200 students each carried out by the researchers. The survey examined the maintenance activities according to: (1) breakdown maintenance, (2) routine maintenance (performance based) and (3) preventive maintenance [46,47]. The institutions were sampled according to their size, data was gathered by a trained surveyor and the information was collected from the city database and from the site survey. The data collected included: floor area (sq.m.), number of students, density of students (students/sq.m.), BRI score and MI score.

b. Gathering of the data of safety hazards in the institutions in all disciplines of the buildings' systems; the data was gathered through interviews with the facilities' maintenance managers using the BRI (Building Risk Indicator) described below;

c. Assessment of the maintenance performance and safety of the facilities using a 25-point rating scale of the severity and probability of failures; assessments were carried out by the researchers through walk out and detailed site surveys in the facilities;

d. Case study and validation of the proposed framework in a public facility using the safety (BRI) and maintenance performance (MI—Maintenance Indicator) models.

The research used 25-point rating scale for the assessment of the safety and maintenance performance of the facilities. The safety assessment and the maintenance performance scales are presented in Tables 2 and 3, respectively, and the deterioration and safety assessment criteria are described in Appendix A. The scales follow the outline developed by

Ni et al. [48] for safety assessment and were further developed for deterioration assessment. The deterioration categories definitions based on the 5-point rating scale introduced in Shohet [19,46]. The safety assessment categories examine the safety consequences, and the 25-point scale represents the risk potential as a multiplication of the severity (consequences) of hazards by the likelihood (probability). These definitions were further developed in this study to 24 categories of core safety components and systems as follows: Infrastructures, Yard organization, Ground leveling, Walking trails, Fence, Gates, Parking, Sports facilities, Preventive measures, Fire Protection, Escape Preparations, Nuisance, Windows and bars, Doors, Railings, Stairs and stairways, Structure, Electric Panels, Lighting, and Emergency lighting and electric wirings. These categories were selected from the review of the literature and are the guide for the safety of educational institutions in the literature review as key maintenance performance and safety performance indicators [11,12,49]. A Building Risk Indicator (BRI) of 0–4 indicates low risk, i.e., safe facility (within this category, between 0 and 2 represents exceptionally safe and between 2 snda4 safe). Building Risk Indicator of 5–12 indicates moderate to marginal and unacceptable risk; within this category, 5–8 represents moderate and unacceptable risk, and a BRI of 8–12 represents marginal-severe risk. The BRI of 12–25 indicates risk at highly severe and up to critical level (this category requires immediate repair of the faults and closure of the institution due to regulatory restrictions). The Building Risk Indicator refers to architectural design, building services schemes safety, the surrounding environment safety, operations and maintenance safety and to safety management activities as derived from Ho et al. [41]. In a similar manner, a maintenance indicator (MI) of 0–4 indicates a durable and well-maintained facility, and within this category, a maintenance indicator between 0 and 2 represents a high durability, and maintenance indicator of 2–4 indicates light deterioration due to aging of the component. A maintenance indicator of 5–12 indicates moderate deterioration (5–8) and marginal maintenance performance (8–12), and between 13–25 indicates highly deteriorated system and lack of maintenance.

Table 2. Grading scale of safety risks of educational buildings systems.

Severity \ Probability	Rare 1 0–10%	Low 2 11–40%	Moderate 3 41–60%	High 4 61–90%	Frequent 5 91–100%
Critical 5	5	10	15	20	25
Severe 4	4	8	12	16	20
Moderate 3	3	6	9	12	15
Minor 2	2	4	6	8	10
Negligible 1	1	2	3	4	5
Legend			High Risk Moderate Risk Low Risk		

This stage laid the ground of the risk informed framework with safety risk and maintenance assessment criteria. These will be used to examine the correlation between the two core variables and as tools for implementing the framework.

Table 3. Grading scale of maintenance of educational buildings systems.

Severity / Probability	Rare 1 0–10%	Low 2 11–40%	Moderate 3 41–60%	High 4 61–90%	Frequent 5 91–100%
Critical 5	5	10	15	20	25
Severe 4	4	8	12	16	20
Moderate 3	3	6	9	12	15
Minor 2	2	4	6	8	10
Negligible 1	1	2	3	4	5
Legend		High Risk	Moderate Risk	Low Risk	

5. Results

The field survey encompassed 24 educational institutes out of 80 in the city of Beer Sheva, composing 30% of the educational facilities population of the city. The survey included background data of the buildings, such as floor area and number of users, as well as assessment of the facilities' maintenance and safety according to the maintenance indicator (MI) and Building Risk Indicator (BRI) defined in the research methodology. The field survey was carried out using a questionnaire and in situ data gathering using a form with detailed scales of safety and maintenance rating criteria. Appendix A presents an example of the rating criteria for exterior cladding safety and maintenance performance. Table 4 summarizes the sample population characteristic background data.

Table 4. Parameters of educational institutes' samples.

Variable	Total	Mean
Floor area (sq.m.)	98,270	4095
Number of students	14,381	599
Density (Number of students/100 sq.m.)	-	15.94
Mean Maintenance score (MI)	-	5.88
Mean Safety Score (BRI)	-	5.42

The mean maintenance indicator (MI) and Building Risk Indicator (BRI) were found to be 5.88 and 5.42, respectively, indicating moderate/marginal maintenance performance and safety risks. The main safety and maintenance faults found are sports grounds, gates, emergency lights and electricity end fixtures. Examination of the correlation between the density of students and the maintenance, as well as the safety performance of the institutions found no systematic ratio, and led to rejection of the second research hypothesis.

5.1. Safety

The highest mean scores of safety risk indicators for the different sections found to be: windows and bars (8.2), static and dynamic nuisances (7.5), wirings (7.0), hallways (6.9), doors and railings (6.7) (Figure 4). Figures 5 and 6 demonstrate examples of typical safety hazards in sports ground and exterior cladding, respectively: lack of safety fixtures and poor maintenance of sport ground (Figure 5a,b, respectively) and deteriorated and poorly maintained exterior stone (Figure 6a,b, respectively).

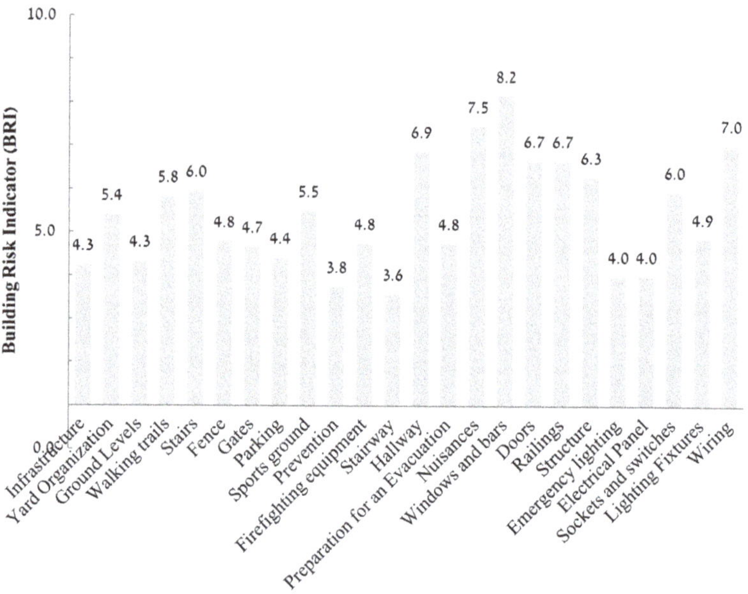

Figure 4. Mean Building Risk Indicator (BRI) in educational facilities sample according to building systems.

(**a**) (**b**)

Figure 5. Sports—(**a**) unprotected basket column and (**b**) poorly maintained basketball surface.

5.2. Maintenance

The following systems found with the highest maintenance score (i.e., most deteriorated): (1) emergency lighting (7.3); (2) static and dynamic nuisances (7.3); (3) electric sockets and switches (7.0); (4) railings (6.9) and (5) gates, preparation for escape and sports grounds (6.8) (Figure 7). The findings demonstrate that electric components, static and dynamic nuisances are the most deteriorated components. Considering MI < 5 as a threshold of high maintenance, it is evident that most of the systems do not meet this performance criterion.

(a)   (b)

Figure 6. (**a**) Deteriorated exterior stone cladding. (**b**) Cracked stone claddings.

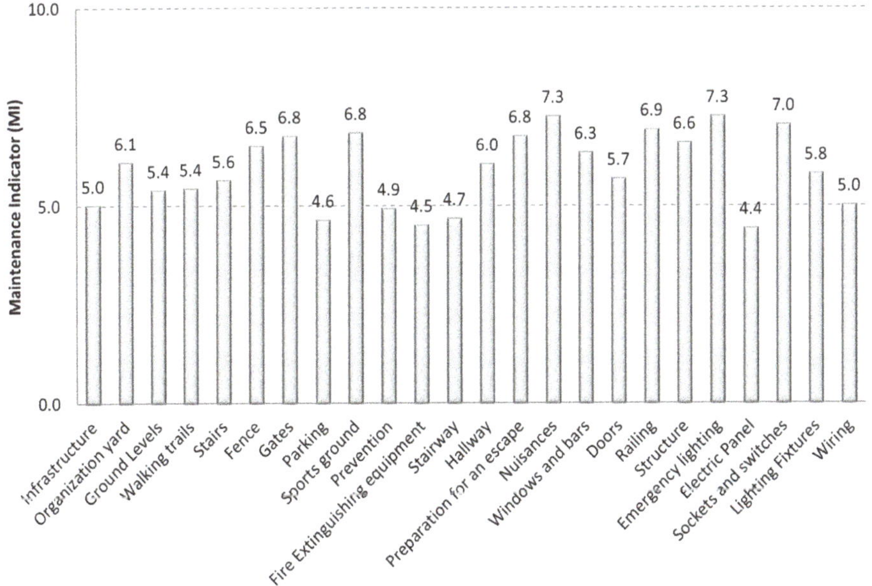

Figure 7. Mean maintenance indicator (MI) for different building and infrastructure systems—Educational Institutions.

5.3. Synergy between Maintenance and Safety

The synergy between maintenance and safety was examined using the Pearson correlation for two variables sample with 22 DOF (Degrees Of Freedom) and at a significance level of 0.01. The critical value for positive correlation is 0.472. Components with Pearson coefficient higher than 0.472 defined as components with strong synergy and those with the Pearson's coefficient lower than 0.472 defined as components with low or no synergy. Figure 8 depicts the distribution of the maintenance indicator (MI) and building risk indicator (BRI) for the 24 systems investigated. The distribution indicates fair link between the safety and maintenance of structure, railings, static and dynamic nuisance, hallways, fire protection, parking, stairs, walking trails, yard organization and infrastructures. The Pearson coefficient for the overall maintenance coefficient found to be 0.7408 indicating that a variance in the maintenance performance explains 74.08% of the variance in safety

(BRI). Table 5 depicts the details of the Pearson's coefficients between the BRI and MI for 24 variables examined within the study. One can deduce that the safety of electric panels and emergency lighting fully explained by the maintenance of these components, 90% of the variance in electric end fixtures, 78% of the variance in parking safety, 49% of the variance in sports facilities and 48% in Fire Protection safety are explained by the conduct of adequate maintenance activities. The findings provide solid evidence of the synergy between safety and maintenance and indicate that 74% of the variance in maintenance is explained by the variance in safety of the facility. Figures 5 and 6 demonstrate a couple of examples of the impact of lack of maintenance of a sports facility (6) and exterior stone cladding (5) on the safety of the facility. It stems from the study that other root causes such as: (1) service regime, (2) design parameters, and (3) safety control and assurance affect the rest of the variance in the facilities' safety risk.

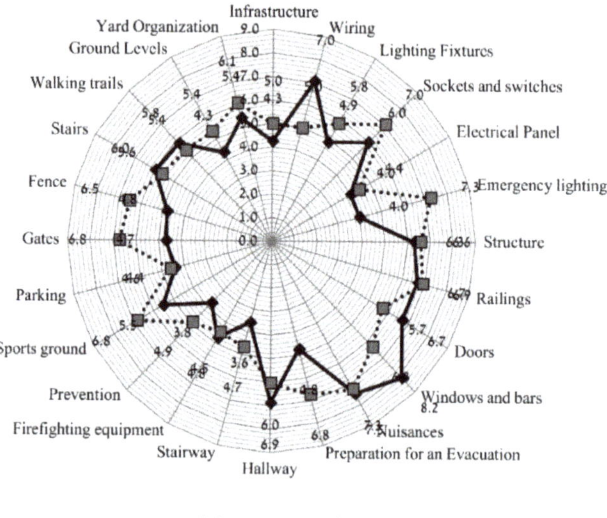

Figure 8. Mean Safety Indicators Vs. Mean Maintenance Indicators—Educational Institutions.

Table 5. Pearson coefficient for the facilities' components and systems.

Component	R	R^2	% of Explained Variance
Infrastructures	0.310	0.10	9.61
Yard organization	0.603	0.36	36.36
Ground leveling	0.422	0.18	17.81
Walking Trails	0.649	0.42	42.12
Stairs	0.114	0.01	1.30
Fence	0.538	0.29	28.94
Gates	0.318	0.10	10.11
Parking	0.886	0.78	78.50
Sports Facilities	0.702	0.49	49.28
Prevention	0.331	0.11	10.96
Fire Protection	0.697	0.49	48.58
Stairways	0.734	0.54	53.88
Hallways	0.444	0.20	19.71

Table 5. *Cont.*

Component	R	R^2	% of Explained Variance
Escape Preparation	0.576	0.33	33.18
Nuisance	0.056	0.00	0.31
Windows and Bars	0.117	0.01	1.37
Doors	0.426	0.18	18.15
Railing	0.475	0.23	22.56
Structure	0.666	0.44	44.36
Emergency Lighting	1	1	100
Electric Panels	1	1	100
Electric Fixtures and Switches	0.952	0.91	90.63
Lighting fixtures	0.334	0.11	11.16
Wirings	0.291	0.08	8.47

Figures 4–8 indicate marginal maintenance and safety performance in the educational facilities sample. It stems from the mean scores of MI and BRI of most of the components at levels higher than 5.74% of the variance in Risk were explained by the execution of maintenance. These findings are demonstrated by Figures 5 and 6. In the next phase of the research a case study of the proposed framework was implemented in public facility to assess the validity of the framework.

5.4. Case-Study—Public Building Integrated Maintenance and Safety

The case study is a public building (courthouse) located in the north of Israel. Its floor area is 26,000 sq.m. of office floors, 3000 sq.m. of parking floors and 4000 sq.m. of gardens and paved access paths. It was constructed in 1999, and its average occupancy is: 340 employees and 3000 visitors per working day (250 days/year). The service regime in the building is standard [50,51]. The integrated risk-informed safety–maintenance framework was implemented in the building between 2010 and 2021 using the safety and maintenance indicators through integrated maintenance–safety audits. An overview of the building, its exterior envelope and electromechanical systems conditions as per 2021 are presented herein. Figures 9–12 illustrate the overview of the building, its exterior envelope, roofing and peripheral infrastructures.

Figure 9. Overview of the building.

Figure 10. Exterior envelope (Maintenance Indicator = 2.5).

Figure 11. Roofing and installations (Maintenance Indicator = 2.5).

The building went under an integrated maintenance–safety performance model according to the framework put forward in this research. The maintenance and safety indicators of the building along the case study implementation period are presented in Figure 13. A high level of fitness between the safety and maintenance performance indicators is observed. Furthermore, consistent improvement in maintenance and safety is observed along the period of implementation. The maintenance and the safety performance of the building as of 2021 (22 years after construction) is remarkably high (2.673 and 2.43 for the maintenance and safety indicators, respectively), The latter indicates a safe working environment, effective safety climate and high-performance building environment.

Figure 12. Stairs and drainage infrastructures around the building (Maintenance Indicator = 2.75).

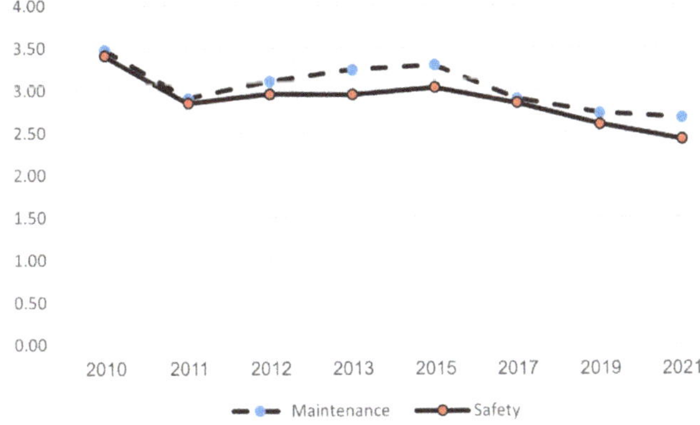

Figure 13. Maintenance and safety indicators in public building case study 2010–2021.

Analysis of variance of the correlation between the maintenance and safety indicators in the case study found a correlation (R^2) of 0.8865 and the p-Value < 0.000478 for significance level of 0.05 (Figure 14). The regression analysis equation is: Y = 0.9135·X + 0.4081 (where Y is the maintenance indicator and X is the safety indicator). The case study strongly validates the safety–maintenance synergy hypothesis resulting from the integrated safety and maintenance performance model proposed in this research. 88.65% of the variance in maintenance performance was explained by the variance in the safety of the facility. The case study demonstrates a strong synergy between the implementation of safety and maintenance models, and that this synergy enhances simultaneous and continuous improvement of the safety, maintenance and performance of buildings. The findings were found to be significant at high level of significance (p Value = 0.00047). These findings support the practical implication of the framework for facility managers.

Regression Analysis	
R	0.941539914
R squared	0.886497409
Adjusted R squared	0.86758031
Standard deviation	0.104030175
Number of samples	8
Coefficient	0.91346406
Standard Erro	0.133438144
T-Test	6.845599263
p-value	0.000477837
Lower limit 95%	0.586952684
Upper limit 95%	1.239975435
Coefficient	0.91346406

ANOVA

	Degree of freedom	SS	MS	F	Level of significance
(Regression	1	0.507156044	0.507156044	46.86222927	0.000477837
Residual	6	0.064933664	0.010822277		
Sum	7	0.572089708			

Figure 14. Regression model and analysis.

6. Discussion

Maintenance and safety of facilities are two close disciplines that share common tools and principles [18–20]. While facilities maintenance is a discipline that carries out preventive, performance-based and breakdown maintenance activities aimed at maintaining the building performance, value and its designated use at high standards, safety discipline uses audits and control tools to maintain acceptable level of risk and maintain the facility safety performance and the wellbeing of the facility occupants.

An integrated theoretical and practical multi-layered risk informed safety–maintenance framework is introduced. The model follows four layers: (1) risk assessment, (2) corrective safety maintenance, (3) maintenance performance assessment and (4) risk-informed preventive maintenance and safety activities. Frangopol et al. [52] introduced the integration of risk, sustainability and resilience into the service life planning of structural bridge systems and components. The model integrates risk and life-cycle loss assessment with multi-objective optimization of bridge and bridge network management. This research emphasizes that a parallel model for educational and other public facilities attains the potential for optimum life cycle costs, as well as meeting the safety and performance constraints.

The preliminary phases of the study revealed that the conduct of high-quality maintenance activities explains 74% of the safety risks variations in educational facilities. The density of the occupants (service regime) does not significantly affect the level of safety risks. The outcomes of the research are similar to preceding research by Arunraj and Maiti [53] in the chemical industry, Geng et al. [54] in virtual maintenance in improving the maintenance safety design and by Khalil et al. [23] in high education facilities. A case study in public facility for a period of 11 years yielded a consistent and continuous improvement of safety and performance of the facility and showed a correlation coefficient of 0.8665 between the safety and the maintenance performance of the facility. The research proposes a novel integrated safety–maintenance management model, and provides solid, validated evidence regarding the synergy between safety and maintenance. The proposed framework may be implemented in various types of facilities such as critical facilities, energy, healthcare and off-shore facilities.

7. Conclusions

The research introduces the topic of synergy and integrated risk-informed safety–maintenance in facilities management. An integrated safety–maintenance model for public facilities was developed and implemented in a case study in the last phase of the research. The following conclusions summarize the research findings:

a. Safety performance in educational institutions was found to be highly dependent on the variance in maintenance performance; this conclusion was deduced from the high Pearson coefficient between the Building Risk Indicator and the Maintenance Indicator;

b. It was found that maintenance activities strongly affect the safety of electric system components (electric panels, lighting, end-fixtures and switches); infrastructures (parking lots, sports facilities, walking trails, and yard organization), fire protection and structural components (stairways, walls, roofs and columns). This conclusion deduced from the partial Pearson coefficients of the facilities' systems between the safety and the maintenance performance for these systems;

c. The study indicates that systematic maintenance of the critical facilities' components such as electric system components, structural components, fire protection and infrastructures implemented with robust, integrated safety–maintenance procedures. This conclusion stems from the inherent dependency between these systems safety and maintenance performance as depicted by the Pearson coefficients discussed above.

d. Annual safety audits of the systems seem to be insufficient in light of the study findings, higher frequencies of maintenance, and safety audits with intervals of between 3 and 6 months are suggested. This conclusion drawn from the marginal performance of maintenance indicator (MI) and safety (BRI) in the sample population, which accomplished in annual safety and performance audits regime.

e. An integrated safety–maintenance performance framework was introduced for synergetic safety–maintenance monitoring, control and management. The framework proposes a cycle loop of safety–maintenance–performance of facilities as a key tool for advanced and effective maintenance and safety management with intervals between 3 and 6 months in public facilities.

f. The framework was validated in a case study of public facility along a period of 11 years. The time history of maintenance performance and safety shows a high level of fitness ($R^2 = 0.8865$ $p < 0.05$). The latter finding supports the practical implication of the framework for facility management applications.

g. This research findings stresses that integrated safety and maintenance should be implemented as a unified and integrated procedure and that this procedure will enhance advanced maintenance performance and safety.

The research develops and provides a robust framework for risk-informed integrated safety–maintenance management framework for educational and public facilities. The theory has been validated in educational and public facilities.

A methodology for the implementation of the framework has been put forward and lay the ground for integrated safety–maintenance methodology. The case study provided solid evidence of sustainable development of the performance of the facility under the implementation of the framework. Future research is recommended to elaborate the framework for automation withing the IFC schema and implement the principles through AI and machine learning tools.

8. Limitations of the Research

The research was carried out on a limited sample of educational public facilities. Extending the research sample could improve the reliability and validity of the research.

Further case studies can elaborate the significance and applicability of the methodology in critical facilities; healthcare and infrastructures such as water, communications, etc.

The research did not refer to the cost-effectiveness of the proposed framework. This should include labor, equipment and spare parts and overheads costs. Analysis of the cost-effectiveness could shed light on the sustainability of the methodology and provide economic and organizational drivers to implement the methodology.

Author Contributions: Conceptualization, I.M.S.; Data curation, R.A.; Investigation, K.-C.W., R.A. and H.-H.W.; Methodology, K.-C.W. and I.M.S.; Project administration, I.M.S.; Software, H.-H.W.; Supervision, I.M.S.; Validation, I.M.S.; Visualization, K.-C.W., R.A. and H.-H.W.; Writing—original draft, K.-C.W. and Writing—review and editing, I.M.S. All authors have read and agreed to the published version of the manuscript.

Funding: This research received no external funding.

Data Availability Statement: The researchers committed to confidentiality of the data collected at educational and public institutions.

Acknowledgments: The authors wish to acknowledge the municipality of Beer Sheva for the availability of the educational facilities data, the courthouse management for accessibility to the case study data and David Ben-Porat for his professional escort of the case study and remarks on the case study.

Conflicts of Interest: The authors declare no conflict of interest.

Appendix A. Static and Dynamic Nuisance Maintenance and Safety Grading Criteria

Table A1. Maintenance grading criteria (MI).

1	2	3	4	5
Cladding is complete and undamaged. No cladding elements have fallen off. Some capillary cracking may be present.	Capillary cracks have developed on portions of the cladding. Single cladding elements have fallen off.	Cracks 0.5 mm wide cover less than 5% of the total cladding area. Up to 3 % of cladding elements have fallen off	Cracks wider than 1 mm have developed on 5% or more of the cladding area. Portions of stone cladding have fallen off.	Significant portions of the cladding have peeled or fallen off. Cracks wider than 5 mm have developed

Table A2. Safety grading criteria (BRI).

1	2	3	4	5
Interior Claddings: Complete, stable, no signs of erosion, degradation, No mechanical deflections, cladding is planar Exterior cladding: Cladding is complete, no minor cracking, no mechanical deformations exterior element properly fixed to claddings.	No erosions, but minor sporadic cracks or spalling in interior cladding, Exterior cladding: Cladding is complete, nor fractures, minor cracks and spalling.	Detachment of up to 3% of cladding, erosion and deterioration of cladding due to intensive use. Fixtures attached to cladding are loosely fixed.	Exterior cladding cracking of up to 5 mm. development of spalling and detachments. Fixtured detached from cladding.	Significant part of cladding have peeled or fallen off. Significant spalling, cladding detached. Cracks wider than 5 mm have developed. Cladding is unstable.

References

1. Achuthan, K.; Murali, S.S. A comparative study of educational laboratories from cost & learning effectiveness perspective. *Adv. Intell. Syst. Comput.* **2015**, *349*, 143–153. [CrossRef]
2. Al-Hemoud, A.M.; Al-Asfoor, M.M. A Behavior Based Safety Approach at a Kuwait Research Institution. *J. Saf. Res.* **2006**, *37*, 201–206. [CrossRef]
3. Lau, E.; Hou, H.C.; Lai, J.H.; Edwards, D.; Chileshe, N. User-centric analytic approach to evaluate the performance of sports facilities: A study of swimming pools. *J. Build. Eng.* **2021**, *44*, 102951. [CrossRef]

4. Cheng, M.Y.; Fang, Y.C.; Chiu, Y.F.; Wu, Y.W.; Lin, T.C. Design and Maintenance Information Integration for Concrete Bridge Assessment and Disaster Prevention. *J. Perform. Constr. Facil. ASCE* **2021**, *35*, 04021015. [CrossRef]
5. Okoro, C.S.; Nkambule, M.; Kruger, A. The state of restroom facilities as a measure of cleaning service quality in an educational institution. *J. Corp. Real Estate* **2020**, *23*, 55–68. [CrossRef]
6. Miraglia, S. A data-driven probabilistic model for well integrity management: Case study and model calibration for the Danish sector of North Sea. *J. Struct. Integr. Maint.* **2020**, *5*, 142–153. [CrossRef]
7. Salaheldin, M.H.; Hassanain, M.A.; Hamida, M.B.; Ibrahim, A.M. A code-compliance assessment tool for fire prevention measures in educational facilities. *Int. J. Emerg. Serv.* **2021**, *10*, 412–426. [CrossRef]
8. Heinrich, H.W. *Industrial Accident Prevention: A Scientific Approach*; McGraw Hill: New York, NY, USA, 1941.
9. DePasquale, J.P.; Geller, E.S. Critical success factors for behavior-based safety: A study of twenty industry-wide applications. *J. Saf. Res.* **1999**, *30*, 237–249. [CrossRef]
10. Addington, L.A. Cops and cameras: Public school security as a policy response to Columbine. *Am. Behav. Sci.* **2009**, *52*, 1426–1446. [CrossRef]
11. Schreck, C.J.; Miller, J.M. Sources of fear of crime at school. *J. Sch. Violence* **2003**, *2*, 57–79. [CrossRef]
12. Schreck, C.J.; Miller, J.M.; Gibson, C. Trouble in the school yard: A study of the risk factors of victimization at school. *Crime Delinq.* **2003**, *49*, 460–484. [CrossRef]
13. Sabatino, S.; Frangopol, D.M.; Dong, Y. Life cycle utility-informed maintenance planning based on lifetime functions: Optimum balancing of cost, failure consequences, and performance benefit. *Struct. Infrastruct. Eng.* **2015**, *12*, 830–847. [CrossRef]
14. Lai, J.H.K.; Yik, F.W.H. An analytical method to evaluate facility management services for residential buildings. *Build. Environ.* **2011**, *46*, 165–175. [CrossRef]
15. Olanrewaju, A.L.; Khamidi, M.F.; Idrus, A. Quantitative analysis of defects in Malaysian university buildings: Providers, Perspective. *J. Retail. Leis. Prop.* **2010**, *9*, 137–149. [CrossRef]
16. Hobbs, A.; Williamson, A. *Aircraft Maintenance Safety Survey: Results*; AR-2007-053; Australian Transport Safety Bureau: Canberra, Australia, 2000.
17. Farrington-Darby, T.; Pickup, L.; Wilson, J.R. Safety culture in railway maintenance. *Saf. Sci.* **2005**, *43*, 39–60. [CrossRef]
18. Hollnagel, E. *Safety Management—Looking Back or Looking Forward, Resilience Engineering Perspectives*; CRC Press: Aldershot, UK, 2008.
19. Shohet, I.M. Key Performance Indicators for Strategic healthcare facilities maintenance. *J. Constr. Eng. Manag. ASCE* **2006**, *132*, 345–352. [CrossRef]
20. Aven, T. *Foundations of Risk Analysis. A Knowledge and Decision-Oriented Approach*; John Wiley and Sons, Ltd.: Chichester, UK, 2003.
21. Hollnagel, E. Risk + barriers = safety? *Saf. Sci.* **2008**, *46*, 221–229. [CrossRef]
22. Straub, D.; Faber, M.H. Systems effects in generic risk based inspection planning. *J. Offshore Mech. Arct. Eng. ASME* **2004**, *126*, 265–271. [CrossRef]
23. Khalil, N.; Kamaruzzaman, S.N.; Baharum, M.R. Ranking the indicators of building performance and the users' risk via Analytical Hierarchy Process (AHP): Case of Malaysia. *Ecol. Indic.* **2016**, *71*, 567–576. [CrossRef]
24. Khan, F.I.; Sadiq, R.; Haddara, M. Risk-based inspection and maintenance (RBIM) multi-attribute decision making with aggregate risk analysis. *Process Saf. Environ. Prot.* **2004**, *82*, 398–411. [CrossRef]
25. Backlund, F.; Hannu, J. Can we make maintenance decisions on risk analysis results? *J. Qual. Maint. Eng.* **2000**, *8*, 77–91. [CrossRef]
26. Khan, F.I.; Haddara, M.R. Risk-based maintenance (RBM): A quantitative approach for maintenance/inspection scheduling and planning. *J. Loss Prev. Process Ind.* **2003**, *16*, 561–573. [CrossRef]
27. De Silva, N.; Malik, R. Maintainability risks of condominiums in Sri Lanka. *J. Financ. Manag. Prop. Constr.* **2010**, *15*, 41–60. [CrossRef]
28. Ruparathna, R.; Hewage, K.; Sadiq, R. Multi-period maintenance planning for public buildings: A risk based approach for climate conscious operation. *J. Clean. Prod.* **2018**, *170*, 1338–1353. [CrossRef]
29. Chiu, C.K.; Chien, W.Y.; Noguchi, T. Risk-based life-cycle maintenance strategies for corroded reinforced concrete buildings located in the region with high seismic hazard. *Struct. Infrastruct. Eng.* **2012**, *8*, 1108–1122. [CrossRef]
30. Wu, Y.; Maravelias, C.T.; Wenzel, M.J.; ElBsat, M.N.; Turney, R.T. Predictive maintenance scheduling optimization of building heating, ventilation, and air conditioning systems. *Energy Build.* **2021**, *231*, 110487. [CrossRef]
31. Duan, C.; Li, Z.; Liu, F. Condition-based maintenance for ship pumps subject to competing risks under stochastic maintenance quality. *Ocean. Eng.* **2020**, *218*, 108180. [CrossRef]
32. Kim, S.; Ge, B.; Frangopol, D.M. Optimum Target Reliability Determination for Efficient Service Life Management of Bridge Networks. *J. Bridge Eng. ASCE* **2020**, *25*, 04020087. [CrossRef]
33. Akanmu, A.A.; Olayiwola, J.; Olatunji, O.A. Automated checking of building component accessibility for maintenance. *Autom. Constr.* **2020**, *114*, 103196. [CrossRef]
34. Park, C.H.; Lim, H. A parametric approach to integer linear fractional programming: Newton's and Hybrid-Newton methods for an optimal road maintenance problem. *Eur. J. Oper. Res.* **2014**, *289*, 1030–1039. [CrossRef]
35. Chen, W.; Chen, K.; Cheng, J.C.; Wang, Q.; Gan, V.J. BIM-based framework for automatic scheduling of facility maintenance work orders. *Autom. Constr.* **2018**, *91*, 15–30. [CrossRef]

36. Chen, C.; Tang, L. BIM-based integrated management workflow design for schedule and cost planning of building fabric maintenance. *Autom. Constr.* **2019**, *107*, 102944. [CrossRef]
37. Saqib, M.; Farooqui, R.U.; Lodi, S.H. Assessment of critical success factors for construction projects in Pakistan. In Proceedings of the First International Conference on Construction in Developing Countries, Karachi, Pakistan, 4–5 August 2008; pp. 392–404.
38. Au-Yong, C.P. Maintenance priority in high-rise housings: Practitioners' perspective versus actual practice. *J. Eng. Res.* **2019**, *7*, 167–177. [CrossRef]
39. Au-Yong, C.P.; Chua, S.J.L.; Ali, A.S.; Tucker, M. Optimising maintenance cost by prioritising maintenance of facilities services in residential buildings. *Eng. Constr. Archit. Manag.* **2019**, *26*, 1593–1607. [CrossRef]
40. Ho, D.C.W.; Chau, K.W.; Cheung, A.K.C.; Yau, Y.; Wong, S.K.; Leung, H.F.; Lau, S.S.Y.; Wong, W.S. A survey of the health and safety conditions of apartment buildings in Hong Kong. *Build. Environ.* **2008**, *43*, 764–775. [CrossRef]
41. Hopkins, A. Thinking about process safety indicators. *Saf. Sci.* **2009**, *47*, 460–465. [CrossRef]
42. Harms-Ringdahl, L. Dimensions in safety indicators. *Saf. Sci.* **2009**, *47*, 481–482. [CrossRef]
43. Ryczyński, J.; Saska, P.; Surowiecki, A.; Ksiądzyna, K. Selected safety issues in designing engineering structures. *Sci. J. Mil. Univ. Land Forces* **2020**, *195*, 135–153. [CrossRef]
44. Lu, Y.; Gong, P.; Tang, Y.; Sun, S.; Li, Q. BIM-integrated construction safety risk assessment at the design stage of building projects. *Autom. Constr.* **2021**, *124*, 103553. [CrossRef]
45. *ISO/DIS 45001:2016*; Occupational Health and Safety Management Systems: Requirements with Guidance for Use. ISO: Geneva, Switzerland, 2016.
46. Shohet, I.M. Building evaluation methodology for setting priorities in hospital buildings. *Constr. Manag. Econ.* **2003**, *21*, 681–692. [CrossRef]
47. Shohet, I.M.; Lavy-Leibovich, S.; Bar-On, D. Integrated maintenance monitoring of hospital buildings. *Constr. Manag. Econ.* **2003**, *21*, 219–228. [CrossRef]
48. Ni, H.; Chen, A.; Chen, N. Some extensions on risk matrix approach. *Saf. Sci.* **2010**, *48*, 1269–1278. [CrossRef]
49. Ministry of Education. Director General Guidelines Paper 5-52: Safety and Security in Educational Institutions. State of Israel. 2013. Available online: http://cms.education.gov.il/EducationCMS/Applications/Mankal/EtsMedorim/5/5-1/HoraotKeva/K-2013-6-1-5-1-52.htm (accessed on 13 April 2019).
50. Shohet, I.M.; Nobili, L. Performance-Based Maintenance of Public Facilities: Principles and Implementation in Courthouses. *J. Perform. Constr. Facil. ASCE* **2016**, *30*, 04015086. [CrossRef]
51. Shohet, I.M.; Nobili, L. Enterprise Resource Planning System for Performance-Based Maintenance of Healthcare Clinics. *Autom. Constr.* **2016**, *65*, 33–41. [CrossRef]
52. Frangopol, D.M.; Dong, Y.; Sabatino, S. Bridge life-cycle performance and cost: Analysis, prediction, optimisation and decision-making. Struct. *Infrastruct. Eng.* **2017**, *13*, 1239–1257. [CrossRef]
53. Arunraj, N.S.; Maiti, J. Risk-based maintenance techniques and applications. *J. Hazard. Mater.* **2007**, *142*, 653–661. [CrossRef]
54. Geng, J.; Zhou, D.; Lv, C.; Wang, Z.L. A modeling approach for maintenance safety evaluation in a virtual maintenance environment. *Comput.-Aided Des.* **2013**, *45*, 937–949. [CrossRef]

MDPI

St. Alban-Anlage 66

4052 Basel

Switzerland

Tel. +41 61 683 77 34

Fax +41 61 302 89 18

www.mdpi.com

Buildings Editorial Office

E-mail: buildings@mdpi.com

www.mdpi.com/journal/buildings

9 783036 562353